D1765102

Web Intelligence XI - CBT
The Ultimate Training Course for Web Intelligence XI

Published by
Schmidt Ink, Inc.
San Diego, CA 92122

Printed in USA

Schmidt Ink, Inc.
Phone: (858) 546-4968
www.schmidtink.com

Other Schmidt Ink, Inc. Publications:

Creating Documents with BusinessObjects
Creating Documents with BusinessObjects: Report Writing Course
Creating Documents with BusinessObjects: Web Intelligence Course
Creating Documents with BusinessObjects: Desktop Intelligence Course
Creating Documents with BusinessObjects XI: Desktop Intelligence Course
Creating Documents with BusinessObjects XI: Web Intelligence Course

Limits of Liability and Disclaimer of Warranty

Trademarks:

ISBN 0-9722636-4-0 (978-0-9722636-4-1)

I dedicate this book to my family. I also dedicate this book to all who believe that regardless of their backgrounds and education levels that they can rise up and learn what others have made seem so far away and difficult. Those that believe that if others can do it, then why can't they. Those that look at a new subject and say, "let's get stated."

BusinessObjects[tm] Web Intelligence XI CBT

Table of Contents

Query Filters - Java Report Panel 37

Working with Report Structures - Java Report Panel 63

Sorts, Filters, Ranks, and Alerters — 169

Query Techniques - Java Report Panel — 193

Thank you for purchasing the Web Intelligence XI CBT. This is one of the most contemporary instructional courses for learning how to use Web Intelligence XI to create documents. It is also the most unique, because it focuses on many real-life examples and situations.

To do the examples in the course the data and universe will have to be configured for use within your WEB Intelligence system. This configuration will have to be completed by a Business Objects professional that is managing your system.

a. This course is supported by the SIEQUITY universe and database, which are found on the System Files directory on the CD.
b. The database that is on the CD is a MS Access database, which should be imported into your database server. Please make sure that the table names remain the same as in the Access database. (If you like, you could just copy the Access database to the WEB Intelligence server.)
c. After importing the data to a database (or copying it to the server), create an OBDC data source with a system DSN named SIEQUITY. Create this data source on the WEB Intelligence server. The universe will look for this DSN.
d. Open the universe in Designer and export it to the Universe repository. If you plan on editing the universe or examining the data in the tables, then you will have to create an ODBC data source, on your local machine, that points to the SIEQUITY data.
e. After the database is imported, the ODBC data source is created, and the universe is exported, then you will be able to use WEB Intelligence to perform the examples in the course.

You may not have access to a Web Intelligence server, if this is the case, please email me at RSchmidt@SchmidtInk.com and I may be able to schedule some time for you on my server. There may be a small fee to cover expenses.

I hope that this training course allows you to reach your Web Intelligence XI reporting goals. It has been carefully designed to allow you to reach your goals and, perhaps, even exceed them. Please have patience and try to relax, as there is a lot of information to assimilate.

If you need help with a problem, please email me at rschmidt@schmidtink.com. You can also visit http://www.forumtopics.com/busobj/viewforum.php?f=35. This is a great web site, where you can have many of your questions answered.

Good luck, and please let me know your comments on the course, as it is written with you in mind.

Very Sincerely,
Robert D. Schmidt

Creating Documents with Web Intelligence XI

Course Introduction

Introduction

- This is a course on Creating Documents with Web Intelligence. We will use the Java Report Panel to create documents. We will also explore the Interactive viewer.
  - We will not cover everything in the Business Objects workspace. However, we will discuss the various options for viewing reports.
- The Business Objects Workspace is highly customizable and the environment in your office may differ from the one shown in this course. However, much of the functionality should be the same.
- In this chapter, we will be introduced to the Business Objects workspace and the Web Intelligence options.

Slide 2

Business Objects is a Web Application that allows several different reporting modules to be used. There can be up to three different applications used to create documents, the available modules depends on the options that your company has selected. The available applications are Desktop Intelligence, Crystal Reports, and Web Intelligence. In this course, we will use Web Intelligence to create reports.

The Business Objects environment allows for reports created with any of the modules to be viewed through InfoView.

Course Recommendations

- This course consists of a course manual and the CBT.
- The CBT acts as a trainer that helps you as you go through the book. The CBT and book are meant to work together to create a very effective learning environment.
- It is best that you have Web Intelligence installed, so that you can work through the examples. I am sorry that I cannot provide you with a demonstration copy, but WEBI is a complex application and needs to be installed by Business Objects professionals.
- The CD contains the database (SI Equity.mdb) and the Universe file for the course. These should be installed by professionals within your company.

Slide 3

I realize that you may not have WEBI installed and available to you while you are studying this book. I have a WEBI server that may be available to you. Please email me at RSchmidt@SchmidtInk.com to schedule time on the server. I cannot promise that this service will be available to you, since there is limited space and licenses. There is also may a fee involved to help cover costs. I make this sever available to you, because it is my goal to make this learning experience as effective as possible.

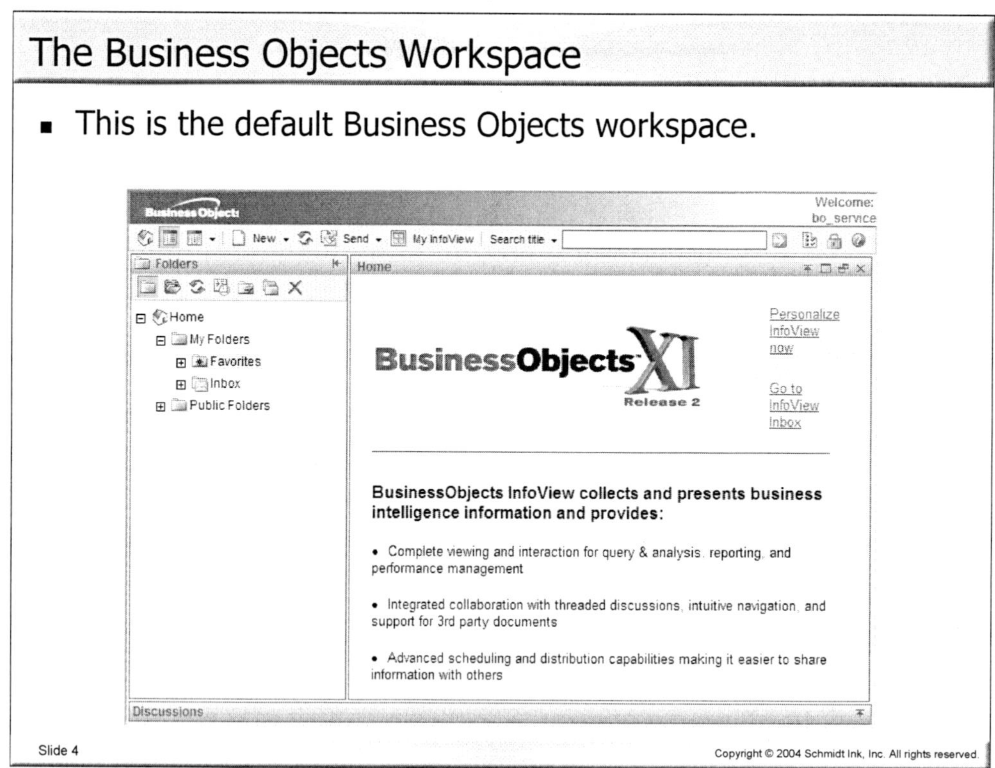

This is the home page of the Business Objects workspace. From here we can Create and View documents. We can have discussion threads. We can also schedule documents to be processed and shared.

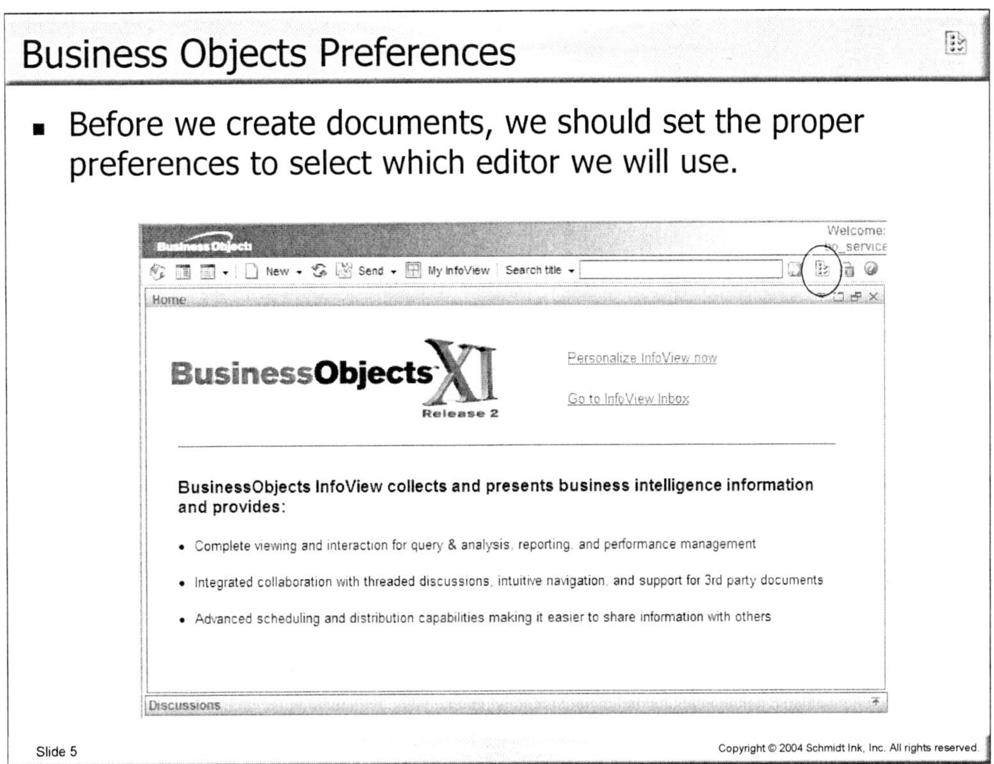

Business Objects Web Intelligence, allows us to use any one of up to three different editors to create documents. Sometimes all three are available, and in other installations, only one is available. Which ones are available depends on how the system is configured and which Web Server that your company is using. In this course, we are going to first use the Java Report Panel. This editor allows us to create the most powerful reports (Version r2).

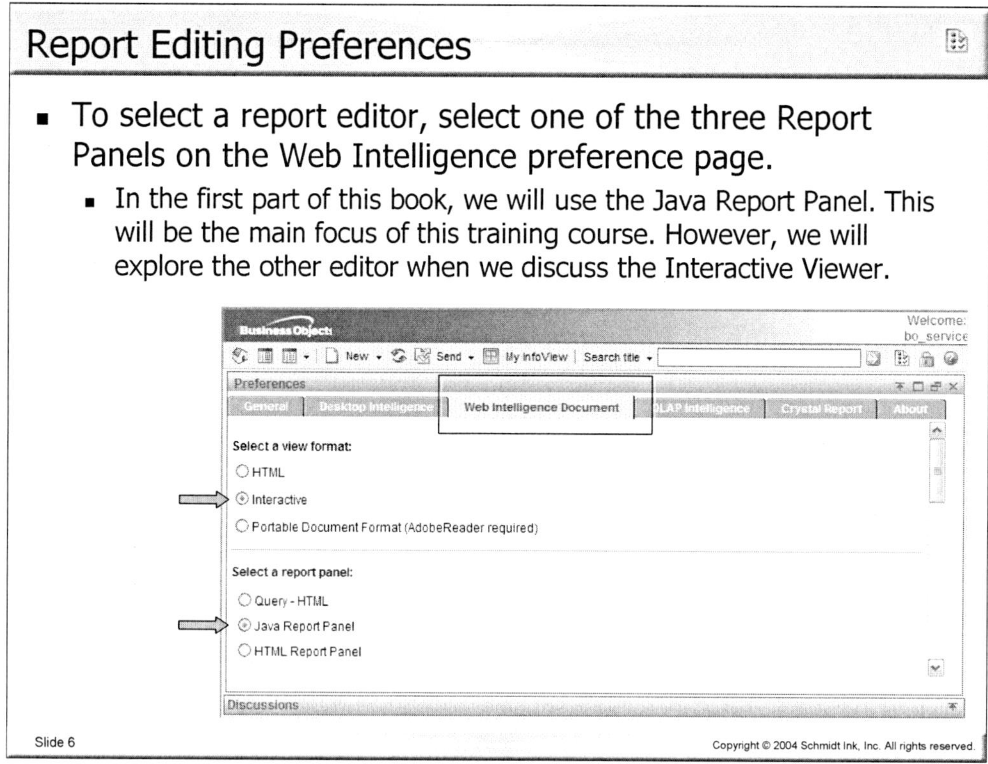

Report Editing Preferences
To select a report editor, select one of the three Report Panels on the Web Intelligence preference page.
In the first part of this book, we will use the Java Report Panel. This will be the main focus of this training course. However, we will explore the other editor when we discuss the Interactive Viewer.
Preferences
Web Intelligence Document
Select a view format:
HTML
Interactive
Portable Document Format (AdobeReader required)
Select a report panel:
Query - HTML
Java Report Panel
HTML Report Panel
Discussions
Slide 6
Copyright © 2004 Schmidt Ink, Inc. All rights reserved.

In this course, we will primarily use the Java Report Panel to create documents. To select this editor as our default editor, click the Preferences button in the Business Objects workspace.

Select the Java Report Panel as Document Editor

1. Click on the Preferences button in the Business Objects workspace.
2. Click on the Web Intelligence tab to activate that section.
3. Select *Interactive* in the *Select a View Format* section. (Not really necessary for the Java Report Panel)
4. Select *Java Report Panel* in the *Select a Report Panel* section.
5. Scroll to the bottom of the page and click the OK button.

Summary

- We are now ready to move to the first chapter of this course – Creating Queries.
  - If we have set the options correctly in this chapter, then we should be using the Java Report Panel.

Slide 7

In this course, we will primarily use the Java Report Panel to create documents. To select this editor as our default editor, click the Preferences button in the Business Objects workspace.

Slide 8

Creating Documents with Web Intelligence XI

Creating Queries
(Java Report Panel)

Introduction

- We are going to start with creating queries, because we will need basic reports to work with throughout the course. By the end of this course, we will create dozens of documents and reports. This should help us to become more confident in our workplace when confronted with the task of creating documents.
- Queries supply a document with data and every document needs data to be useful. In BusinessObjects, we create a query to retrieve data, and then we format the data into useful information.
- In some companies, it is very easy to retrieve data. In others, it can be a bit trickier. The data we use in this course is not super easy, but not as tricky as some.

Slide 10

Many people say data is not useful and cannot be called information, until it is properly formatted to emphasize the impact that the data has on our business. This is probably one of the only general statements that I believe can be applied to most instances. Nobody wants to look at a page of data that is not organized to highlight the purpose of the data. In this chapter and a few of the following, we are going to work on bringing data into a document. Then, later in the course, we are going to discuss how to format this data into useful information.

Put on your seatbelt, because after this course, you will never look at data, information and reports as you have in the past. We are going to learn how to be data analysts and document creators. Not just employees that create reports that people have asked us to create.

BusinessObjects Universes

- In Web Intelligence, we retrieve data for our reports through a Universe. BusinessObjects universes are logical mappings of data in a database.
- Universes are generally targeted towards certain aspects of a business. For example, a company may have a universe for Sales, Inventory, Shipping, Human Relations, and so forth.

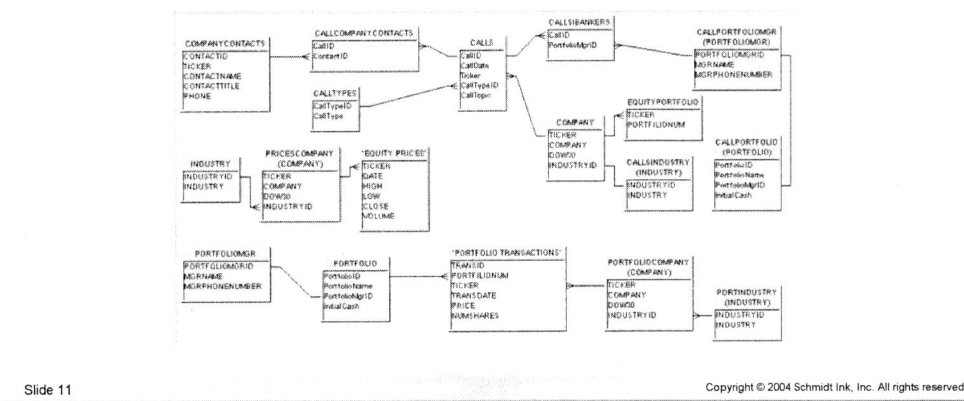

Slide 11

Most report developers do not usually have to create universes. This is usually done by specialists that understand how the data is related and how the data is to be used.

In most cases, you do not have to know the structure of a universe to create documents. However, some universes are very complicated and it behooves us to know how the tables are related in the database. With such knowledge, we can make our queries much more efficient.

If we plan on developing reports for a career, it is also probably a very good idea to learn SQL. It is a good skill and usually beneficial to those trying to become professional report developers.

BusinessObjects Universe Contexts

- Universe contexts define logical groups of tables that cannot be directly related in a universe. In SI Equity there are three contexts
  - Calls, Portfolio, and Equity Prices
- Not all universes contain contexts, but if they do, developers usually identify them with folders, as we have done below.

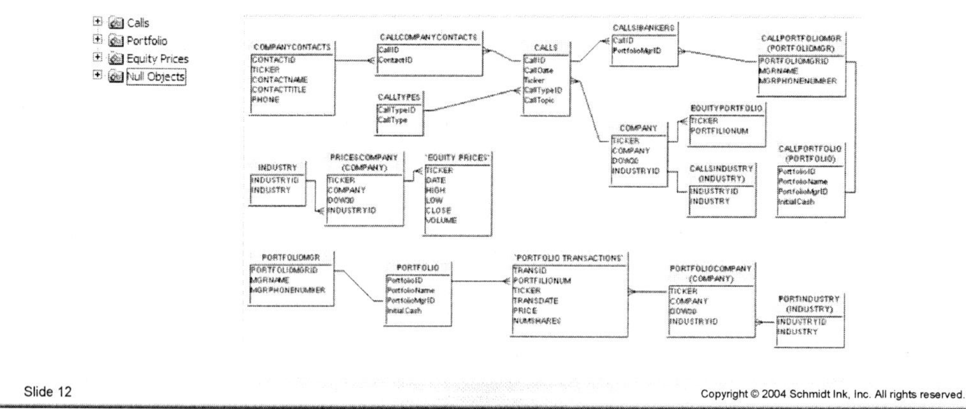

Slide 12

Contexts allow us to have more than one fact table in our universe. A fact table is a table that contains the transactions of a certain aspect of our business. For example, contact calls, marketing, sells, shipping, inventory, equity transactions, and so forth.

We usually can not directly relate two different fact tables, because the relationship may cause the measures in a report to become artificially inflated due to a multiplying effect from the relationship. Therefore, in order to have two different fact tables in a universe, contexts are usually defined to keep the fact tables independent of one another.

In SI Equity, we have three contexts:

- Calls: This context allows us to get information on any meetings that our portfolio managers have had with our clients.
- Portfolio: This context allows us to get all of the information on our portfolios and the transactions that each has made.
- Equity Prices: This is for the price history for the stocks that are represented in our different portfolios.

Unfortunately, we usually can not create queries using objects from more than one context. For example, we can not create a simple query that returns objects from both Calls and Portfolio. However, as we become more advanced BusinessObjects developers, we will learn how to use these different contexts to combined information in a report.

Start a New Web Intelligence Document

- To start a new Web Intelligence Document, select Web Intelligence Document from the New drop-down menu.
- You may see the Java logo as it loads the Web Intelligence application. If this is your first time, you may also be prompted to allow the Java plug-in to be installed.

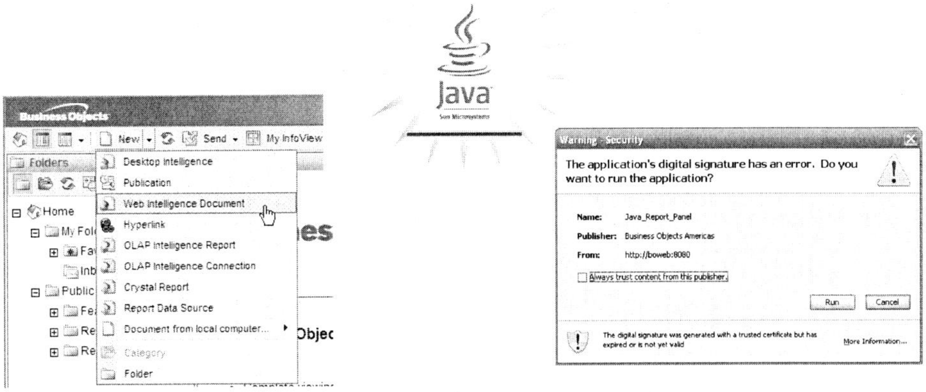

Slide 13

In this manual, I will not tell you how to start InfoView, because many companies have different methods of accessing the application. If you are unsure of how to open InfoView, then please ask your BusinessObjects' manager for instructions.

In the screen above, there are many applications to choose from. You may only have one, several, or even more applications than I.

Start a New Document

1. Start InfoView.
2. Select Web Intelligence Document from the New drop-down menu.
3. If the Warning –Security dialog is displayed. Ask your administrator, if you can click the Run button.

Selecting a Universe

- The first step to creating a Web Intelligence document is to select a universe for the document.
  - This example displays three universes. The top two (eFashion and Island Resorts Marketing) ship with the BusinessObjects application. The SI EQUITY universe is supplied on the CD accompanying this course. We will use this universe for the examples in course.
  - To select a universe, simply click on it to start your document.

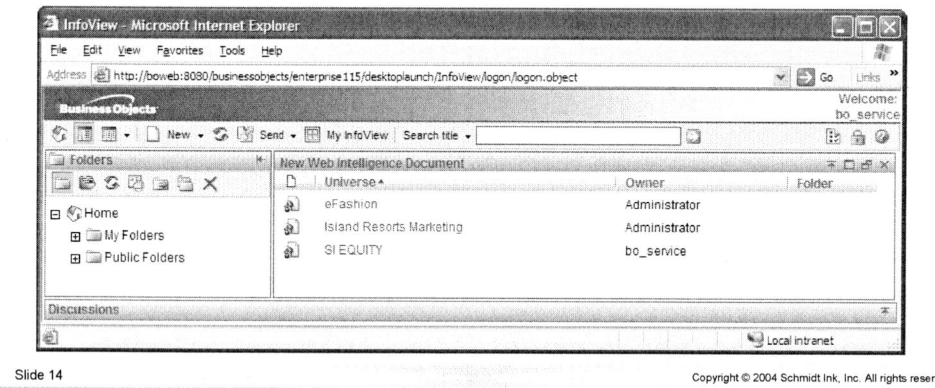

Slide 14

Universes are designed to allow us to access the data needed for our documents. Some departments may only have one or two, and some departments may have dozens. Some universes are very targeted and simple. Others are very convoluted and difficult to navigate. In any case, boldly stepping into a universe is the best policy. You cannot hurt it, so select one and explore what is available.

After selecting a universe, BusinessObjects will pop-up their logo while it loads. You may also be prompted to allow plug-ins to install. Plug-ins are programs that allow applications to work within the Internet Explorer environment.

Select the SI Equity Universe

1. Click on the SI EQUITY universe.

Edit Query Environment

- This is the environment where we create and edit our document's queries. There are three major sections
  - The Data and Properties pane
    - All objects available in the universe are located in this pane. We browse the class folders to find objects of interest.
  - The Result Objects pane
    - This is where we place the objects that are to return information to our document.
  - The Query Filters pane
    - Objects in this pane, define filters that are placed on our queries.

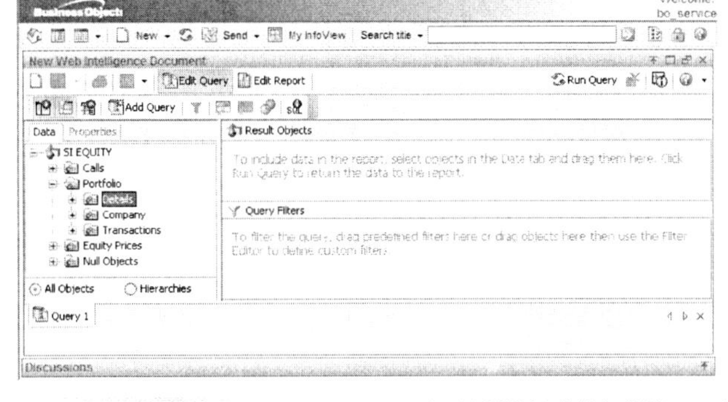

Slide 15

The query environment allows us to define and edit queries in our document. It is very simple to navigate. We use queries to instruct Web Intelligence to retrieve data of interest. To create a query, we simply create combinations of objects from the Data tab in the Result Objects pane.

Data Types

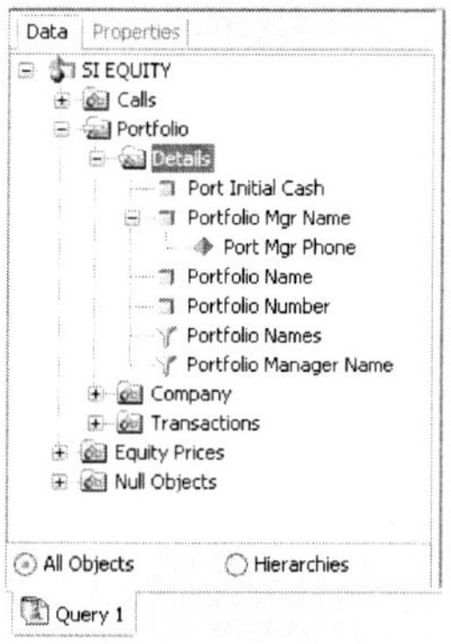

- **Class**
  - Used to organize objects by category or group.
- **Dimension Object**
  - Usually identify information, such as Client ID's, stock ticker symbols, and locations. Since, dimensions identify information, they are used to synchronize data providers, define sections in Master-Detail reports, and define rankings in a report.
- **Detail Object**
  - Usually describe some attribute of a dimension. They usually contain information, such as weight, phone numbers, employee names, and so forth.
- **Measure Object**
  - Usually are aggregates that conform to the dimensions in a report. They will usually sum, count, min, max, or average numerical information to a context defined by dimension values.

Slide 16

It is important to know what the different data provider object types represent and how they will behave in our reports. As we become more advanced, knowledge of these different types will allow us to create more and more powerful reports.

We will talk about the condition types in the next chapter. For right now, we are only interested in the behavior of the objects that will populate our report with data.

Data Tab of the Data Manager

- The Classes and Objects window of the Query Panel contains all of the objects that are available in a universe.
- To open a class and expose the objects.
  - Click on the plus (+) sign preceding the class.
  - Or, Double-click on the class.
- Details are associated with Dimension objects.
  - If a Dimension object is preceded with a plus (+), then there is a Detail associated with it.
  - Details usually have a one-to-one relationship with their parent dimension.
    - This means that for each dimension value, there should only be one detail value.

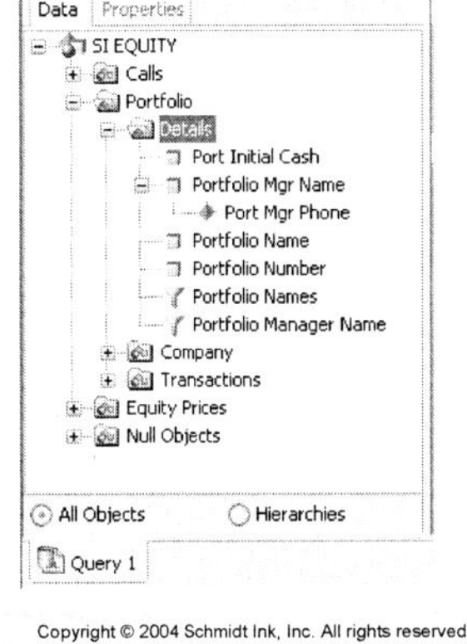

Slide 17

All of the data objects in BusinessObjects are organized into classes – the ultimate class being the name of the universe. Many times classes are just used to help categorize available objects, and objects from any class can be used with objects from other classes. However, this is not always the case. For example, in the SI Equity universe, there are three main classes – Calls, Portfolio, and Equity Prices. Objects from these folders can not be combined with objects from the other classes in the list. For example, if an object from the Calls class is placed in the Result Objects pane, then an object form Portfolio class cannot be placed in the pane. These classes are known as incompatible.

Details are used to associate some attribute to a dimension. For example, a product may have weight, color, or a description. An employee dimension may have a name detail object associated with it. Many people use name as a dimension, but this cannot always be the case, since names do not uniquely identify an employee. For this reason, designers usually make the employee number a dimension and the name of the employee a detail of the employee number dimension.

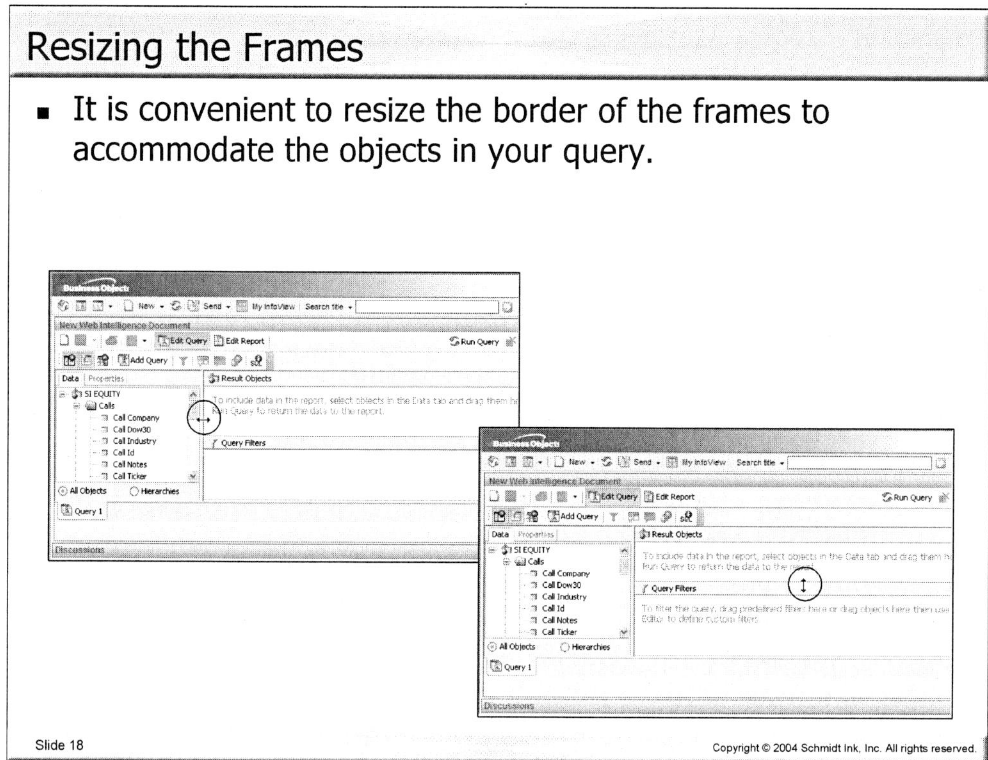

Resizing the Frames
It is convenient to resize the border of the frames to accommodate the objects in your query.
Slide 18
Copyright © 2004 Schmidt Ink, Inc. All rights reserved.

It is important to know that you can resize the frames within the Edit Query Environment. This allows you to see the entire object names in the Data pane and to see as many objects as possible in the Result Objects pane.

Resizing the Panes in the Query Environment

1. Move your cursor over a border in the Edit Query window.
2. When it turns into a drag cursor, click and drag the border to a new width.

Selecting Objects for a Query

- To create a query, we simply place objects from the Data tab into the Result Objects pane. To place objects in the Result Objects window, we can
  - Double-click objects on the Data tab.
  - Drag and drop objects into the Result Objects pane.
    - When dragging objects, they will be placed preceding the object that they are dropped on.
  - Drag and drop entire classes into the window.

Be sure not to select objects from different contexts. If you do, you may get the following error.

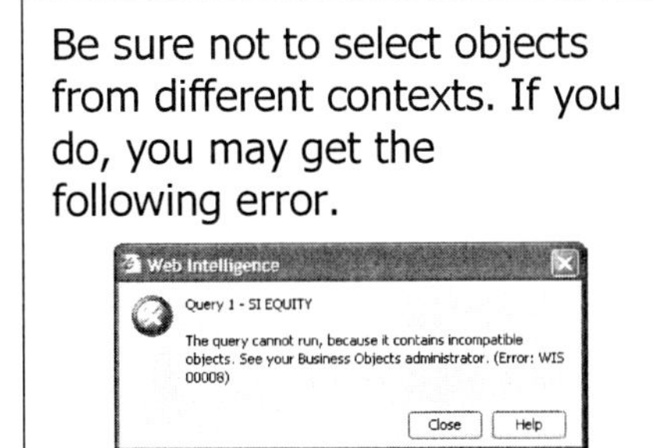

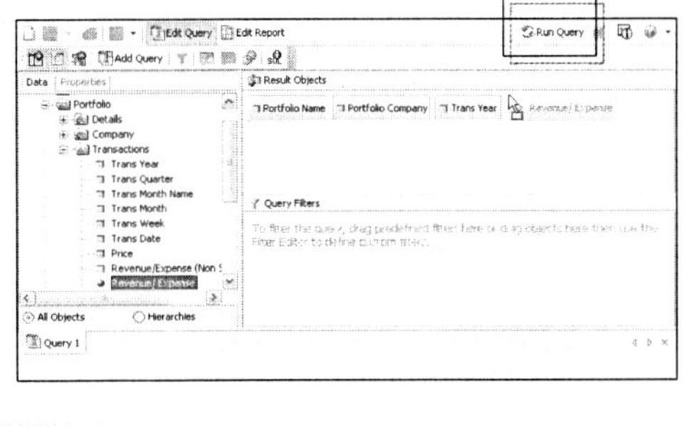

Slide 19

Objects placed in the Result Objects section will define the data set that is to be retrieved by the query. This data set may also define the default table in a new report.

Selecting Objects for a Query

1. Open the Portfolio Class on the Data tab of the Data Manager.
2. Open the Details folder.
3. Drag the Portfolio Name object to the Result Objects window.
4. Open the Company folder.
5. Double-click on the Portfolio Company object.
6. Open the Transactions folder.
7. Double-click on Trans Year and Revenue/Expense.
8. Click the Run Query button.

Query Environment Help Section

- The Help section of the Edit Query Environment displays any description that a universe designer has assigned to an object.
  - Some universes have extensive descriptions, while others have none. It is probably a good idea if your universe has no descriptions, to request descriptions to help eliminate any confusion that may arise about an object's purpose.

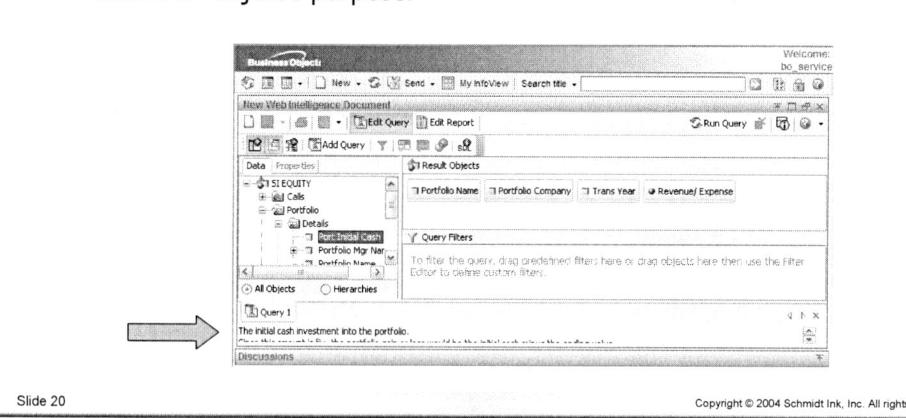

Slide 20

The Help section can be very helpful when trying to determine the purpose for various objects in the panel. However, the universe designers must assign descriptions to the objects in order for there to be help text.

Running the Query

- When creating a new document, Web Intelligence must perform several tasks to complete the request.
- Each task will take a different amount of time.
- The first time a document is created, Web Intelligence will not know how long it may take to execute. However, after a document has been created, Web Intelligence will offer an estimated time based on the previous refresh.

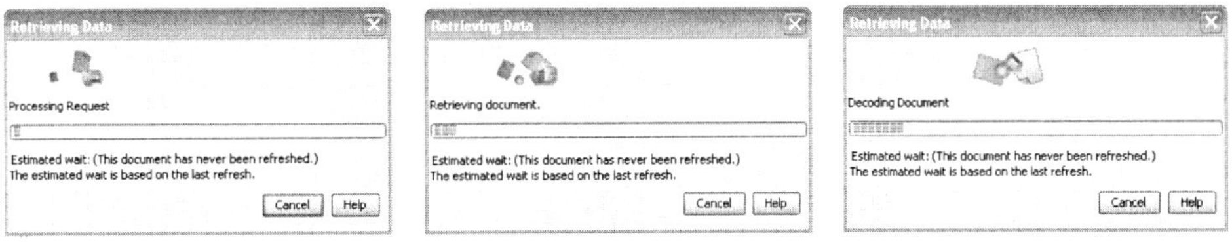

Slide 21

When a document is created with Web Intelligence, most of the processing takes place on the sever. This means that Web Intelligence sends the request to the server, and then waits for the processing to finish. Finally, a finished document is sent back to Web Intelligence to be displayed.

The following are the steps that probably take place on the server...

1. It must create SQL from the objects that you have selected. (SQL are the instructions that are sent to a database to retrieve a data set. We do not have to know how to create SQL, because BusinessObjects will do this for us)
2. We must wait for the database to collect the data. (This could be really quick or painfully slow. It all depends on the complexity of the query and the size of the data set)
3. The database must send the information to server application.
4. Once all of the data is received, the server must calculate all of the formulas in the report. Once this is done, it populates all of the cells in the report with the results. (This phase is usually relatively quick, but it can be slowed down by many complex calculations)
5. Then the web page is constructed and sent to your Web Intelligence.

The Default Report

- When a new document is created, Web Intelligence uses a default template to create a report with all the objects in the Query Results section. Throughout much of this course, we are going to modify and enhance this default table to create powerful reports.
- The column header titles are created from the object names.
    - To change a column header title, double-click on the header and type in a new title.

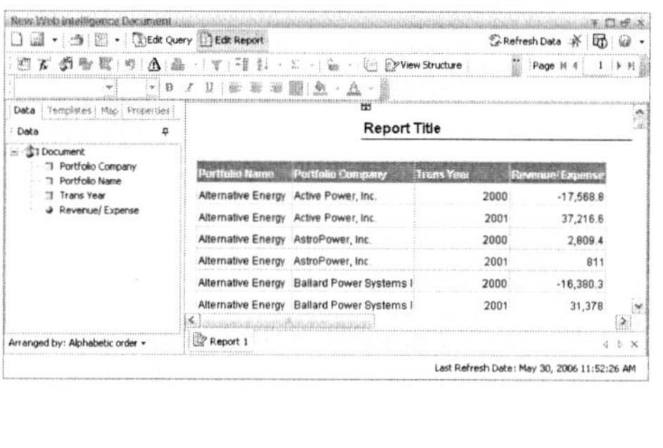

Slide 22

The default report is usually just a table with a report name header. This report is rarely used as is, so in this course we will usually modify the default table into an informative report.

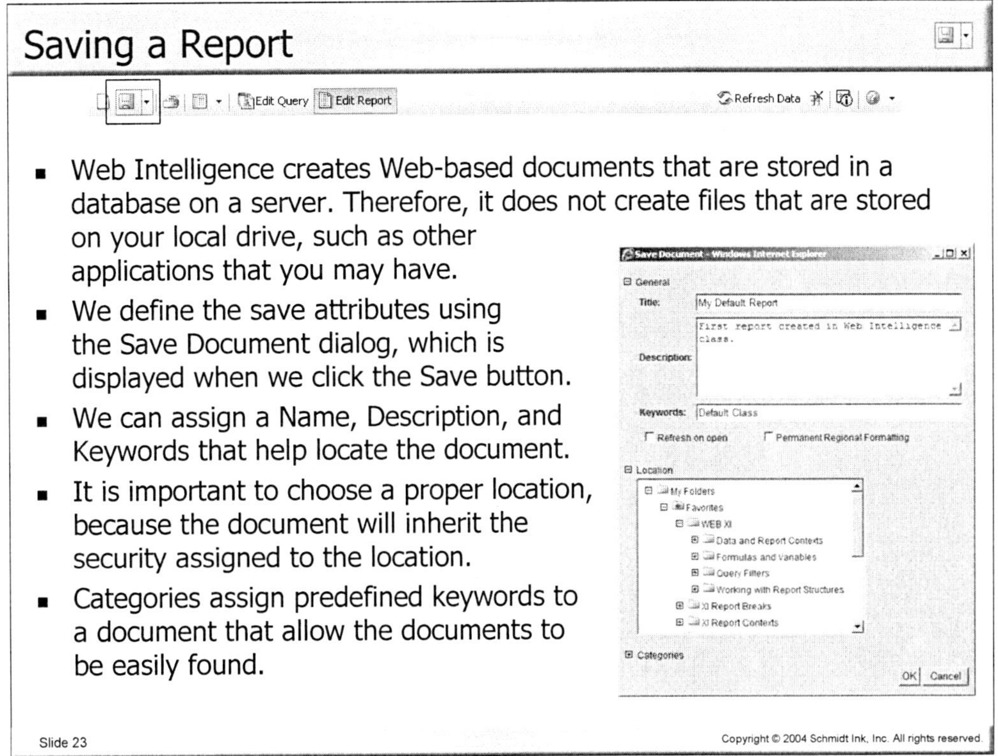

Saving a Report
Edit Query
Edit Report
Refresh Data
Web Intelligence creates Web-based documents that are stored in a database on a server. Therefore, it does not create files that are stored on your local drive, such as other applications that you may have.
We define the save attributes using the Save Document dialog, which is displayed when we click the Save button.
We can assign a Name, Description, and Keywords that help locate the document.
It is important to choose a proper location, because the document will inherit the security assigned to the location.
Categories assign predefined keywords to a document that allow the documents to be easily found.
General
Title:
My Default Report
Description:
First report created in Web Intelligence class.
Keywords:
Default Class
Refresh on open
Permanent Regional Formatting
Location
My Folders
Favorites
WEB XI
Data and Report Contexts
Formulas and variables
Query Filters
Working with Report Structures
XI Report Breaks
XI Report Contexts
Categories
OK
Cancel
Slide 23
Copyright © 2004 Schmidt Ink, Inc. All rights reserved.

Many of us are used to saving documents to a drive on our system. Then, when we want to open them, we simply browse to that directory and open the document. However, Web Intelligence does not save documents to a system drive. It stores them in a secure database that is accessed through the Business Objects InfoView environment. This makes our document secure and accessible from any computer that has access to your Web Intelligence environment.

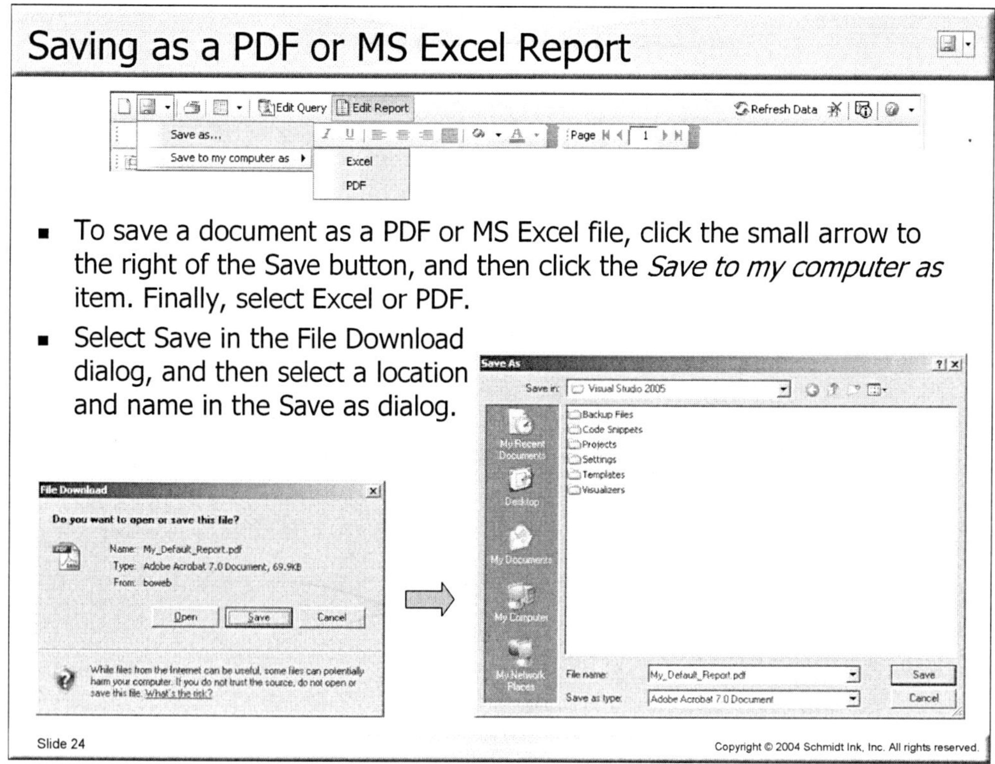

Sometimes we still need a hard file on our system drive that we can email or manipulate in some fashion. To save a document to your file system, select the *Save to my computer as...* menu item. Saving a document as PDF or Excel will cause that copy to loose all Business Objects attributes, therefore you will not be able to edit it in the Web Intelligence environment. However, you will still have the Web Intelligence version, since the *Save to my computer as* option only saves a copy of the original Web Intelligence document.

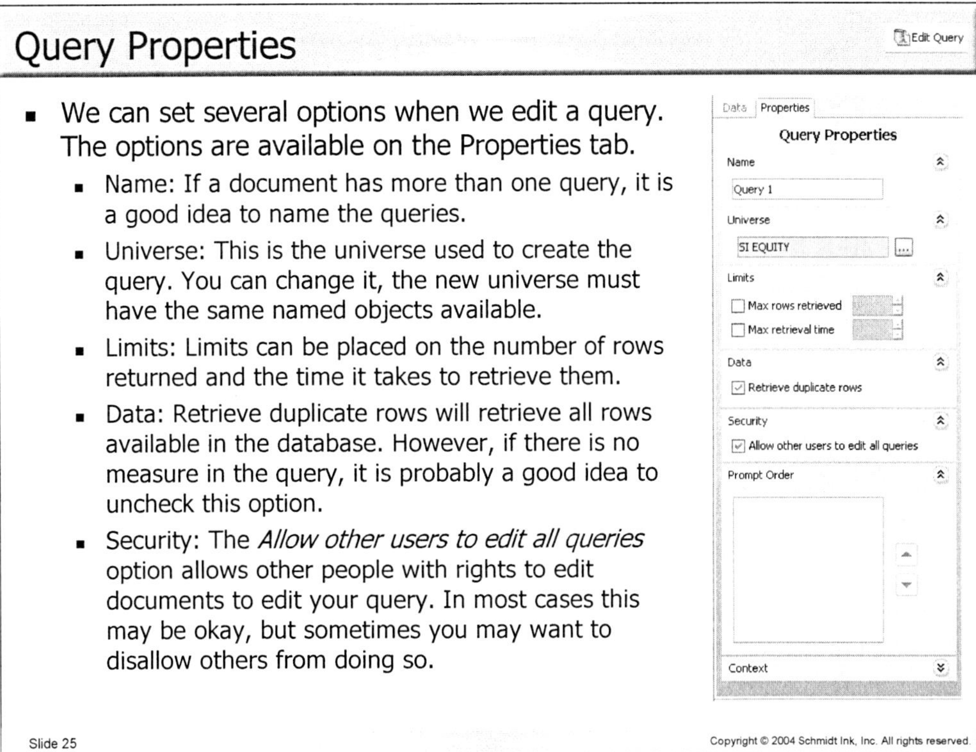

Query properties allow us to set several options for our query. Most of the time, we do not need to worry about these options, but there are times that we should always set some of them.

We should always name our queries, if we have more than one. This will allow others to know the purpose for each query and thus make it easier for others to work on existing documents

We should always uncheck the Retrieve Duplicate Rows option when our query has no measures. Measures cause the query to return unique combinations of dimensions that will have a total associated with them. For example, the total number of sales for each salesperson in each territory.

John	East	10,000
Terri	West	30,000

If the query has no measures, then the dimensions may not return unique combinations and may return many more rows than anticipated. Web Intelligence will roll these rows up on the report and display only unique combinations, even though the query may contain hundreds or even thousands of duplicate rows. This is why it is best to clear this option, if there are no measures in the query.

John	East
John	East
John	East
John	East
Terri	West
Terri	West
Terri	West

Edit Query Toolbar Buttons (Pane Options)

Edit Query — To display the Edit Query environment in an existing document, click the Edit Query button.

- Shows or hides the Data and Properties tab.
- Shows or hides the Filter pane.
- Shows or hides the Scope of Analysis pane.

We will not discuss the functionality all of these options in this chapter. However, we will cover them in later chapters.

Slide 26

The Edit Query environment is very dynamic. We can work with several different panes within the window. We can hide the Data and Properties pane, when it is of no value to the current task. We can show or hide the filter pane. We can display the Scope of Analysis pane to help us define the drilling capability of our document. We can also add another query using the Add Query button. We will talk about each of these tasks as the course goes on.

Edit Query Toolbar Buttons (Query Options)

Edit Query — To display the Edit Query pane in an existing document, click the Edit Query button.

Add Query — Adds another query to the document, by adding another Query tab to the Edit Query environment.

- Places a Quick Filter on an object that is associated with a list of values (Usually a Dimension Object).
- Allows the creation of a subquery. (Discussed in a later chapter)
- Allows the application of a database ranking. Only available for universes of certain databases.
- Used to create combination queries, such as union, intersect, or minus.
- Used to view or modify SQL in a query.
- We will not discuss the functionality all of these options in this chapter. However, we will cover them in later chapters.

Slide 27

The Query Option buttons allow us to modify the logic of our queries. They allow us to develop more complex queries.

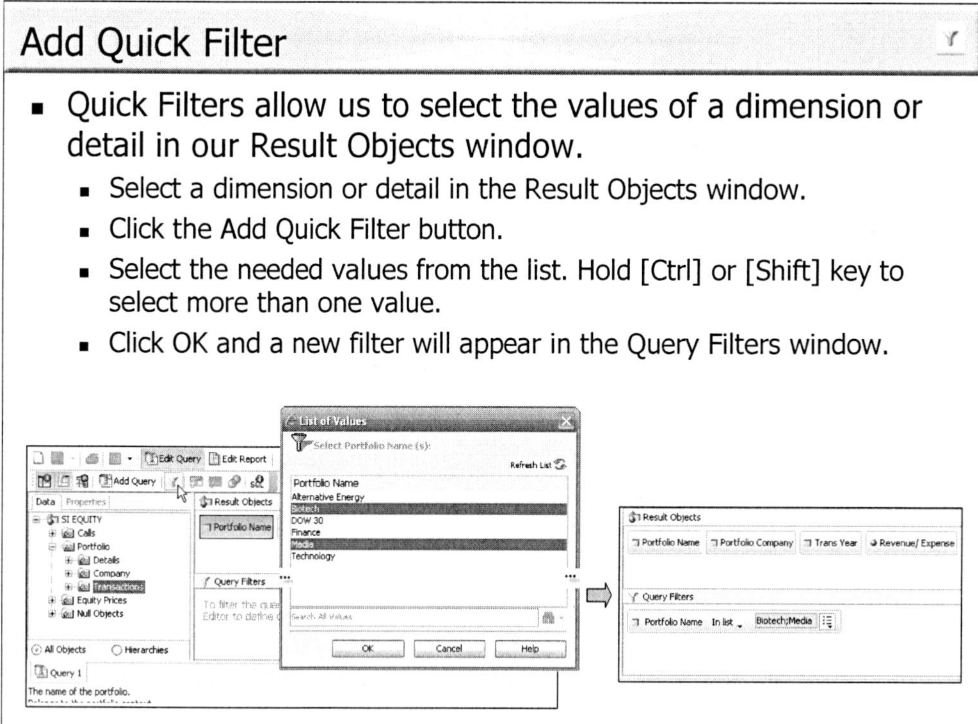
Add Quick Filter
Quick Filters allow us to select the values of a dimension or detail in our Result Objects window.
Select a dimension or detail in the Result Objects window.
Click the Add Quick Filter button.
Select the needed values from the list. Hold [Ctrl] or [Shift] key to select more than one value.
Click OK and a new filter will appear in the Query Filters window.
List of Values
Portfolio Name
Alternative Energy
Biotech
DOW 30
Finance
Media
Technology
OK
Cancel
Help
Result Objects
Portfolio Name
Portfolio Company
Trans Year
Revenue/ Expense
Query Filters
Portfolio Name In list Biotech;Media
Slide 28
Copyright © 2004 Schmidt Ink, Inc. All rights reserved.

We can place quick filters on dimensions and details in the Result Objects section of the panel, only if they have a list of values associated with them. This means that some objects may not be compatible with the Add Quick Filter button. When the object is not compatible, the Add Quick Filter button will be disabled.

Place a Quick Filter on an Object

1. Click the New toolbar button to create a new Web Intelligence document.
2. Select the SI Equity universe.
3. Open the Portfolio class and double-click Portfolio Name, Company Name, Trans Year, and Revenue/ Expense.
4. Click on the Revenue/ Expense object in the Result Objects window.
    - Notice that the Add Quick Filter button is disabled.
5. Click on the Portfolio Name object in the Result Objects window.
6. Click the Add Quick Filter button.
7. Click on Biotech in the list.
8. Hold down the [Ctrl] button and click on Media.
9. Click the OK button.
    - Notice the Filter in the Query Filters window.
10. Click the Run Query button to create the document.
    - The report should have information for both Biotech and Media. Later in this course, we will learn how to format this type of report using breaks or sections.

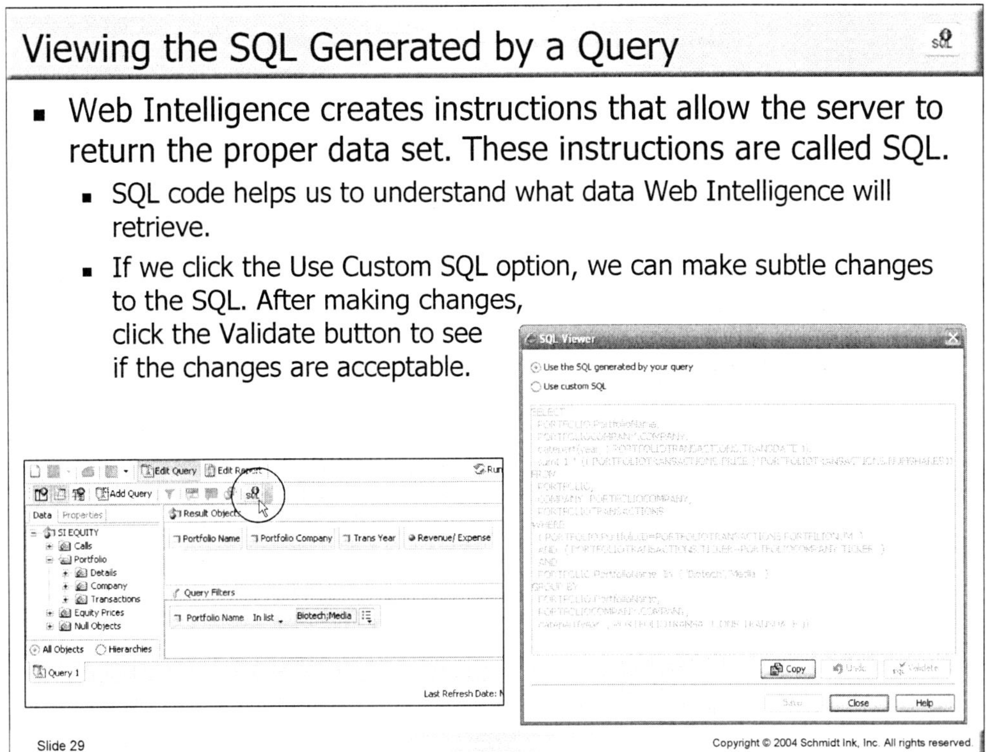
Viewing the SQL Generated by a Query
Web Intelligence creates instructions that allow the server to return the proper data set. These instructions are called SQL.
SQL code helps us to understand what data Web Intelligence will retrieve.
If we click the Use Custom SQL option, we can make subtle changes to the SQL. After making changes, click the Validate button to see if the changes are acceptable.
SQL Viewer
Use the SQL generated by your query
Use custom SQL
Edit Query
Add Query
Result Objects
Portfolio Name
Portfolio Company
Trans Year
Revenue/ Expense
Query Filters
Portfolio Name
In list
Biotech;Media
SI EQUITY
Calls
Portfolio
Details
Company
Transactions
Equity Prices
Null Objects
All Objects
Hierarchies
Query 1
Last Refresh Date:
Copy
Close
Help
Slide 29

When first becoming familiar with a universe, it may be helpful to view the SQL generated by the Query. If you do not understand SQL, don't worry, you will still be able to create reports. The SQL just helps you to understand more.

View SQL of a Query

Click the Edit Query button, after completing the previous exercise.

1. Click on the SQL button.
2. Click OK, when done viewing the SQL.

Add Query

- Web Intelligence allows us to have more than one query in a document. We could use multiple queries to
  - Supply both detail and summary information. We often do this if the sum of the details does not match the sum in the summary. This often happens when departments share the credit for certain measures within the company.
  - Import data from multiple universes or contexts within the same universe. (See example on the next slide)
  - There are other valid reasons that we won't list.
- Sometimes people use more queries than necessary
  - Suppose that you wanted the number of transactions and the amount of sales. You could easily build one query to return this information. However, some developers will build a separate query for both needs.

Slide 30

Being allowed to use more than one query in a document allows us to create more powerful reports. For example, sometimes it is just too complicated to sum the details to get a total in a report. An example might be a company that sales one investment type, such as a short term loan. Let's say the company that borrowed money is Microsoft. Now, Microsoft can be considered a technology company and it can also be considered a media company. Suppose that both the Media and Technology departments in a company claimed the loan in their department's revenue. Let's say the loan was for $400,000. The totals for each department would contain the 400,000.

Now, we want to create a company total. So, we add the details from both the technology and media departments and we get 400,000 more than we should have. After analyzing the data, we find out that we double-counted the 400,000. This is the way our company does business, so we can not tell them that only one department should claim the 400,000. We therefore, use one query for the details, and then create another (without the department dimension) for the company total.

I hope that this does not confuse you, but I wanted to mention it here, because it is so common. We will do an example of this later in the course.

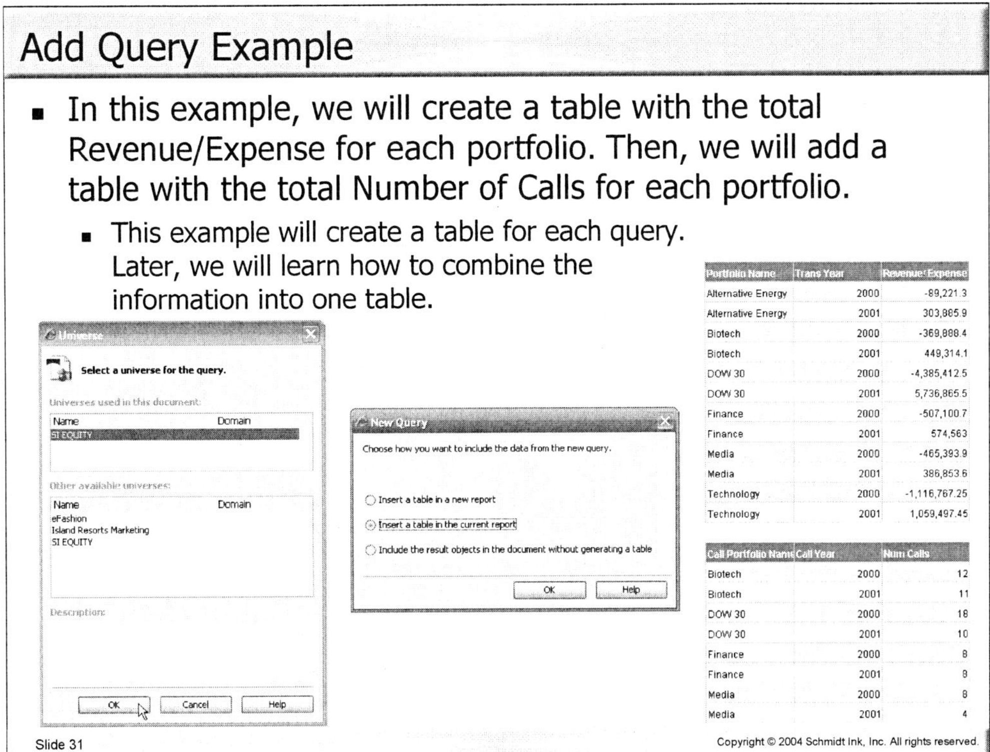

Portfolio Name	Trans Year	Revenue/Expense
Alternative Energy	2000	-89,221.3
Alternative Energy	2001	303,885.9
Biotech	2000	-369,888.4
Biotech	2001	449,314.1
DOW 30	2000	-4,385,412.5
DOW 30	2001	5,736,865.5
Finance	2000	-507,100.7
Finance	2001	574,563
Media	2000	-465,393.9
Media	2001	386,853.6
Technology	2000	-1,116,767.25
Technology	2001	1,059,497.45

Call Portfolio Name	Call Year	Num Calls
Biotech	2000	12
Biotech	2001	11
DOW 30	2000	18
DOW 30	2001	10
Finance	2000	8
Finance	2001	8
Media	2000	8
Media	2001	4

Multiple Query Document

(First table)

1. Click the New toolbar button to create a new Web Intelligence document.
2. Select the SI Equity universe.
3. Open the Portfolio class and double-click Portfolio Name, Trans Year, and Revenue/ Expense.
4. Click the Run Query button.

(Second Table)

5. Click the Edit Query button to return to the Edit Query environment.
6. Click the Add Query button to add a new query.
7. Select the SI EQUITY universe from the Universe list and click the OK button.
8. Open the Multiple class within the Calls class.
9. Double click on Call Portfolio Name
10. Double-click Call Year and Num Calls in the Calls class.
11. Click the Run Queries button.
12. Click the *Insert a table in the Current Report* option, in the New Query prompt dialog.
13. Click the OK button.

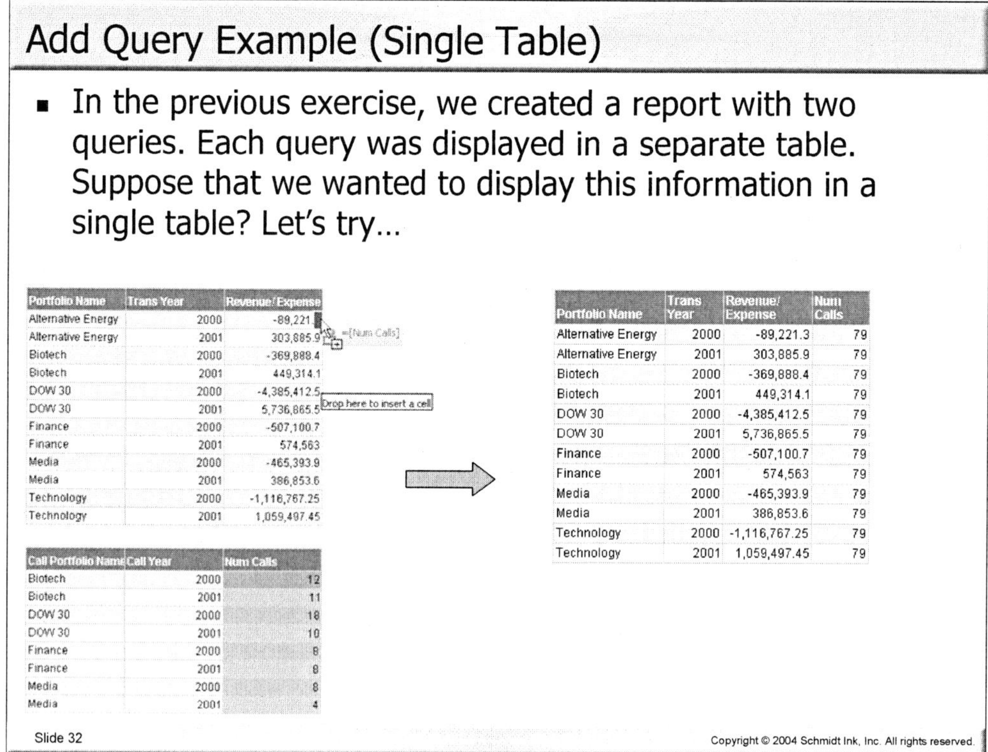
Add Query Example (Single Table)
In the previous exercise, we created a report with two queries. Each query was displayed in a separate table. Suppose that we wanted to display this information in a single table? Let's try...
=[Num Calls]
Drop here to insert a cell
Slide 32
Copyright © 2004 Schmidt Ink, Inc. All rights reserved.

Portfolio Name	Trans Year	Revenue/Expense
Alternative Energy	2000	-89,221.3
Alternative Energy	2001	303,885.9
Biotech	2000	-369,888.4
Biotech	2001	449,314.1
DOW 30	2000	-4,385,412.5
DOW 30	2001	5,736,865.5
Finance	2000	-507,100.7
Finance	2001	574,563
Media	2000	-465,393.9
Media	2001	386,853.6
Technology	2000	-1,116,767.25
Technology	2001	1,059,497.45

Call Portfolio Name	Call Year	Num Calls
Biotech	2000	12
Biotech	2001	11
DOW 30	2000	18
DOW 30	2001	10
Finance	2000	8
Finance	2001	8
Media	2000	8
Media	2001	4

Portfolio Name	Trans Year	Revenue/ Expense	Num Calls
Alternative Energy	2000	-89,221.3	79
Alternative Energy	2001	303,885.9	79
Biotech	2000	-369,888.4	79
Biotech	2001	449,314.1	79
DOW 30	2000	-4,385,412.5	79
DOW 30	2001	5,736,865.5	79
Finance	2000	-507,100.7	79
Finance	2001	574,563	79
Media	2000	-465,393.9	79
Media	2001	386,853.6	79
Technology	2000	-1,116,767.25	79
Technology	2001	1,059,497.45	79

In the above slide, we dragged the Num Calls measure from a table and dropped it into another table. We did this by clicking on any cell in the Num Calls column and then dragging it into the right side of the Revenue/Expense column in the other table. This caused a new column to be inserted and then populated with the Num Calls object. However, the Num Calls object is the same value for every row in the table. Since this is highly unlikely, there must be something that we did not consider.

Multiple Query Document

1. Using the report created in the previous exercise, drag the Num Calls object to the rightmost part of the Revenue/Expense column in the other table. Before you release the mouse button, hold down the [CTRL] key to copy the column and not move it.

Add Query Example (Single Table)

- Two data sets can be related by the common dimensions that they contain.
  - For example, suppose that we have two data sets
    - Debbie has 10,000 shares of IBM
    - Debbie has $50,000
  - When we look at this data, we assume that Debbie has 10,000 shares of IBM and $50,000. However, this is only true if both Debbie's are the same person.
  - In this case Debbie is a dimension, because the name identifies a person. The 10,000 shares and $50,000 are both measures. This means that if we can associate both Debbie's as the same person, then the amounts can simply be assigned to a single row.
  - In BusinessObjects, we associate the Debbie's (Dimensions) by merging the dimensions into a single object.

Slide 33

The above example seems kind of silly, but it is exactly how BusinessObjects works with multiple data sets that are to be related. We simply find the dimension objects in all sets of data that identify the same object. Then, we merge these dimensions into a single object.

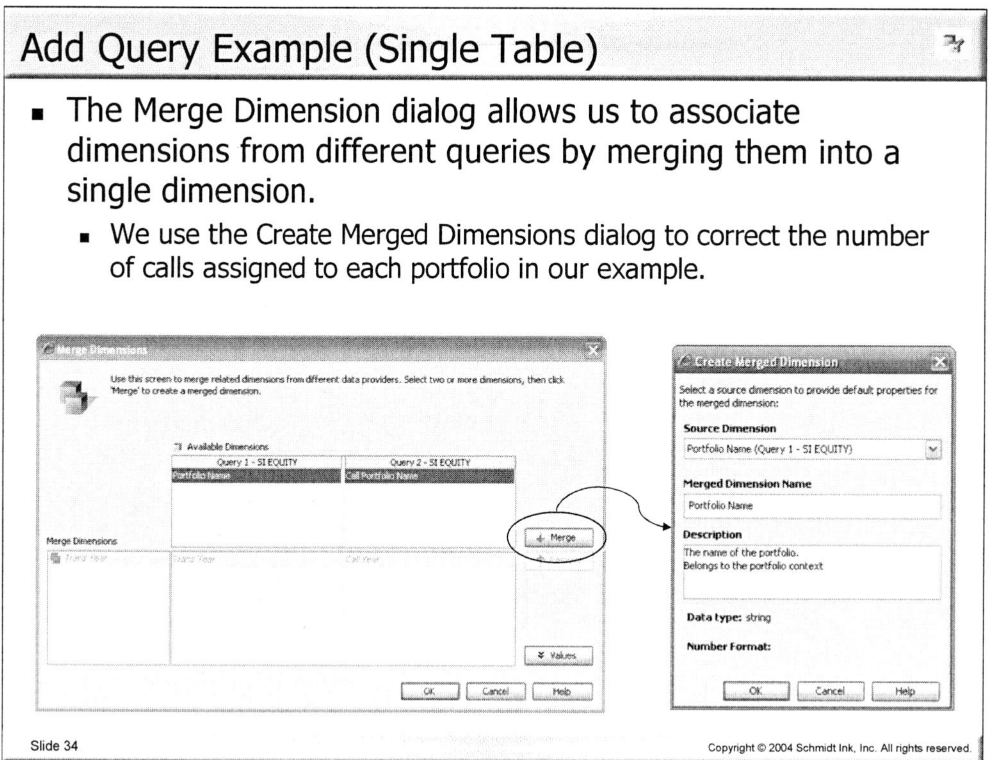

The Merge Dimensions dialog allows us to merge dimensions from different queries into a single dimension in our document. These creations are only available in the document in which they are created and have no effect on other documents within the BusinessObjects system. Once created, we can use the new object as any other object in the document.

Merge Dimensions in our Document

1) Click the Merge Dimensions button to display the Merge Dimensions dialog.
   Notice that the dialog is populated with dimensions from both queries.
2) Click on the Trans Year object in the Available Dimensions list.
   Notice that only Call Year is enabled in the other query (Query 2). BusinessObjects does some checking for compatibility.
3) Click on the Call Year dimension to select it.
   Notice that now that two dimensions are selected, the Merge button has become enabled.
4) Click the Merge button.
   The Create Merged Dimension dialog is displayed. This is where we can name and describe our new object.
5) Click the OK button.
   The new Merged Dimension object is now listed in the Merged Dimensions list.
6) Repeat the above for Portfolio Name and Call Portfolio Name.
7) Click OK to dismiss the Merge Dimensions dialog.

Add Query Example (Single Table)

- If you did the exercise on the previous page, your table should now display the proper number of calls for each Portfolio Name.
- This technology allows us to create many mixed context or universe reports, another example could be
  - The current orders of a product from the sales universe and the number of the product left in inventory from the inventory universe.

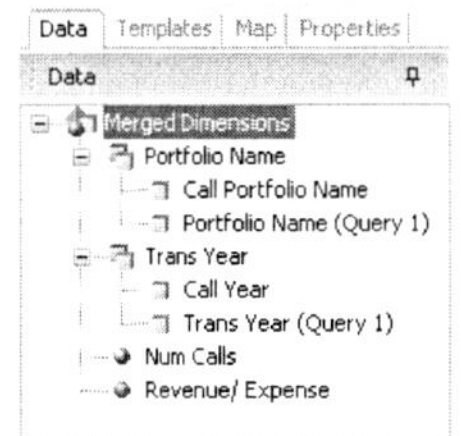

Portfolio Name	Trans Year	Revenue/ Expense	Num Calls
Alternative Energy	2000	-89,221.3	
Alternative Energy	2001	303,885.9	
Biotech	2000	-369,888.4	12
Biotech	2001	449,314.1	11
DOW 30	2000	-4,385,412.5	18
DOW 30	2001	5,736,865.5	10
Finance	2000	-507,100.7	8
Finance	2001	574,563	8
Media	2000	-465,393.9	8
Media	2001	386,853.6	4
Technology	2000	-1,116,767.25	
Technology	2001	1,059,497.45	

- Merged dimension are displayed in the data tab as shown above.

Slide 35

The ability to merge dimensions in a document allows us to relate data from different queries in a document. With this ability, we can create very powerful reports. For example:

- The money spent on marketing in each sales region from the marketing universe and the amount of sales in each sales region from the sales universe.
- The cost of each product from the manufacturing universe and the average selling price from the sales universe.

The list could go on much further since almost every business has the need for such technology.

Creating Queries Summary

- In this chapter
  - We learned how to select objects from a universe.
  - We also learned about the different types of objects that are available.
  - We found out that a universe can contain multiple contexts, which can make some objects incompatible.
  - We also learned how to add an additional query to our document. We were able to display the data in two different tables.
  - We merged the common dimensions in multiple queries so that we could display the measures from two different queries in a single table.

Slide 36

This chapter has been a great introduction to Web Intelligence query development. We have learned many new things and this seems like a good place to end the chapter. It may behoove you to take a short break here. In the next chapter, we are going to discuss how to place query filters on our queries.

Creating Documents with Web Intelligence XI

Query Filters
(Java Report Panel)

Introduction

- Query Filters allow users to retrieve only the information of interest from a database.
  - For example, a database may contain information for many different years or products. Query Filters allow constraints to limit the information retrieved to selected years or products.
- Since query filters limit data in a document, the right combination of query filters can make documents more efficient and responsive.

When users forget to place query filters on queries, they may resort to extreme measures to express their displeasure.

Slide 38

Conditions allow us to retrieve only data of interest from a database, which makes our queries more efficient and compact. We should carefully consider each condition and make sure it is doing exactly what we need it to do. For example, some company's fiscal year is not the same as a calendar year. Therefore, it is important for them to use a fiscal year condition to retrieve data for a certain year.

If we did not have conditions, we would have to bring the entire data set to our document and then filter the data to show only what interests us. This would make our documents awkward and difficult to work with. This example is extreme, but it does illustrate the importance of conditions in a query.

As we get more advanced, we will learn more and more creative ways to place conditions on our queries. We will not cover all of these methods, but it is important to realize that there are many creative ways to create conditions to get exactly what we need from our data providers.

Designer Defined Query Filters

- Universe Designers often provide pre-defined Query Filters, which are represented by yellow funnels in the Data tab.
  - These pre-defined Query Filters are usually very efficient and effective, because the Universe Designer has put an effort to create the most requested filters.
- To apply a predefined Query Filter
  - Locate the desired filter in the classes and objects of the data tab.
  - Double-click on the filter or drag it into the Query Filters window.

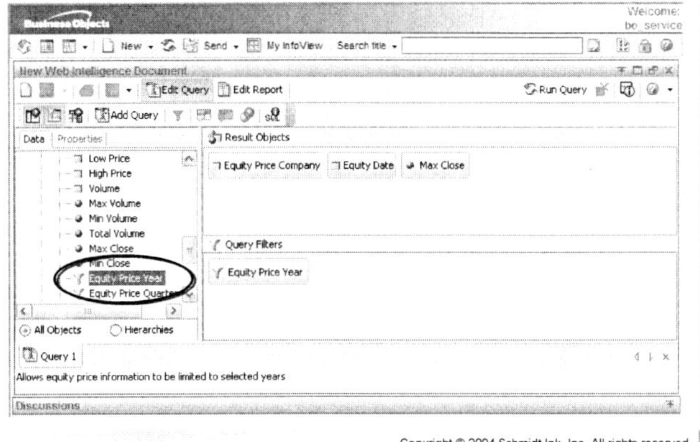

Slide 39

When universe designers create universes, they not only create dimensions, details, and measures, they also should create Query Filters that we can quickly and confidently place on our queries. Many times the filters that they create are more complicated than the ones that we can create on our own within Web Intelligence. It is also an advantage to use designer created filters, because the prompts, if any, will be consistent throughout all of the documents..

Using a Designer Defined Query Filter

1. Select *New Web Intelligence Document* from the menu.
2. Select the SI Equity universe as the data provider.
3. Open the Equity Prices folder.
4. Double-click on Equity Price Company, Equity Date, and Max Close.
5. Double-click on the Equity Price Year Query Filter in the Data tab.
6. Click the *Run Query* button.
7. Enter 2001 in the *Type a Value* field. You may also double-click on it in the *List of Values* section of the dialog.
8. Click Run Query.

The Query Filter that we have selected in this example will prompt us for a year value, when the query is refreshed. Not all Query Filters will prompt for values.

Entering Values for Prompted Queries

- Prompted Query filters allow values for the operand to be entered when a query is refreshed.
- Many filters have a list of values to select from.
- The List of Values displays all values available in the data source.
- To refresh the list, click the Refresh List button.

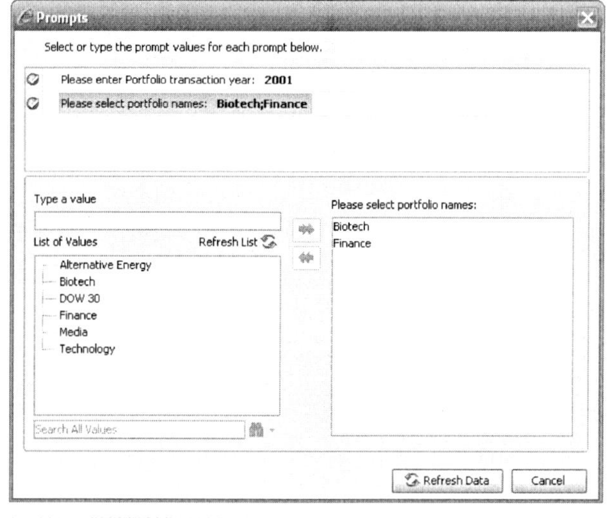

Slide 40

Web Intelligence allows for Query Filters that prompt for the operand value(s) when a query is refreshed. This allows the document to be *customized* when it is refreshed, by allowing the refresher to choose parameters, such as date ranges, departments, products, and so forth. In many cases, a value can be typed into the Type a Value edit field or values can be selected from a list. Usually, operands, such as Department or Region (dimensions), will have a List of Values. Operands, such as revenue or amounts (measures), usually will not have a list.

Create a Prompted Query

1. Select *New Web Intelligence Document* from the menu.
2. Select the SI Equity universe as the data provider.
3. Open the Portfolio folder.
4. Open the Details sub-folder.
5. Double-click on the Portfolio Name dimension and the Portfolio Names query filter.
6. Open the Transactions sub-folder.
7. Double-click on the Trans Year dimension, the measures Revenue/ Expense and Num Transactions, and the query filter Portfolio Transaction Year.
8. Click the *Run Query* button.
9. Select the first prompt in the list – Please enter Portfolio transaction year:.
10. Double-click the 2001 value in the list of values.
    (If there are no values in the list, click the Refresh List button)
11. Select the Second prompt in the list – Please select portfolio names:.
12. Double-click Biotech and Finance in the list of values.
    (If there are no values in the list, click the Refresh List button)
13. Click Run Query to create the report.

The Query Filter that we have selected in this example will prompt us for a year value, when the query is refreshed. Not all Query Filters will prompt for values.

Combining Multiple Query Filters

- Universe Designers often provide pre-defined conditions that can be used to limit data.

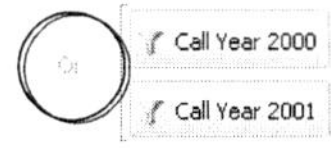

- When combining two or more conditions on the same object, the conditions must be Or'ed together. This example reads - The year can either be 2001 or 2000.

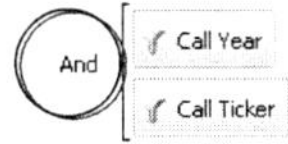

- When combining two or more conditions on different objects, the conditions should be And'ed together. This example reads – A row of data must contain the specified Call Year and Call Ticker values. Both Query Filters will prompt for a value when the query in refreshed.

Note: To change an And to an Or, or vice versa, double-click on the operator...

Slide 41

Most queries contain more than one condition. Therefore, we must know how to combine them to logically return the data of interest. BusinessObjects supplies us with two logical operators to combine conditions – And & Or.

The And states that two object filters on the same row should be true in order to return the row. The Or states that one of two object conditions on the same row should be true. It also states that a single object should be one of several values to return a row.

The statement **Year 2000 And Year 2001** would return no data to the document, because a year cannot be both 2000 and 2001. In order to retrieve the proper data in this case, we would say **Year 2000 Or Year 2001**. This means that the year could be equal to 2000 or 2001.

The statement **Color is Red Or Car is Mustang**, would not return all red mustangs. It would return all red cars (regardless of make) and all mustangs (regardless of color). The document creator would probably want to use **Color is Red And Car is Mustang**. This would return all red mustangs.

Using Multiple Conditions in a Query

1. Select *New Web Intelligence Document* from the menu
2. Double-click Call Year 2000 and Call Year 2001. In the Calls class folder.
3. Double-click on the And operator to change it to an Or operator. (Now the condition combination will return data if any Call Year equals 2000 or 2001)

Combining And and Or Logic

- Many times we need conditions that combine And and Or logic. When combining logic, the *And* operator has precedence over the *Or* operator.
- Once the filters are placed in the Query Filters section, we can rearrange them with the mouse.
  - To create logical groups (Parenthesis), drag a filter to the bottom portion of another filter in the Query Filters section. A logic framework will appear. Then just drop the filter into the framework.
- Logic rules follow Algebra order of precedence.
  - This means that many logic statements can be simplified. For example, we can simply the logic on the bottom of this slide using the distributive rule. Call Dow 30 And (Call Year 2000 or Call Year 2001)

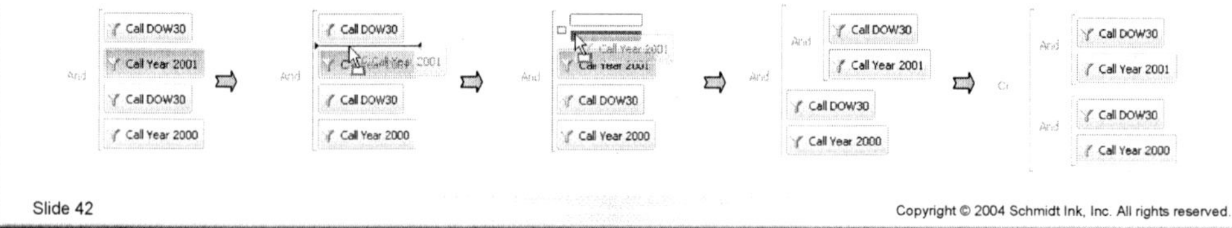

Slide 42

Logic and Algebra have many similarities. For example, the And operator is similar to the multiplication operator and the Or operator is similar to the addition operator. This means that any factoring rules that you learned in Algebra, probably also apply to the logic in your Query Filters window. Most of the time this knowledge is unnecessary, but once in a while it comes in handy.

Combining And and Or Logic

1. Select New Web Intelligence Document
2. Double-click on the following objects in the Calls Class to place them in the Result Objects section
   - Call Portfolio Name in the Multiple class, Call Company, Call Contact Name in the Multiple class, Call Month Number, Call Month Name, and Num Calls
3. Double-click on the following Query Filters in the Calls class
   - Call Dow 30, Call Year 2001, Call Dow 30, Call Year 2000
4. Click on the Call Year 2001 filter (in the Query Filters section) and drag it to the bottom portion of the Call Dow 30 filter.
   - This will cause a logic framework to be displayed. Drop the Call Year 2001 filter in the bottom portion on the logic framework.
5. Click on the Call Year 2000 filter (in the Query Filters section) and drag it to the bottom portion of the Call Dow 30 filter.
   - This will cause a logic framework to be displayed. Drop the Call Year 2000 filter in the bottom portion on the logic framework.
6. Change the outermost And logic operator to an Or operator.
   - Double-click on the outside middle And operator.

Custom Conditions

- Quick filters and designer filters work well in many cases, but sometimes we need to create a Custom Filter.
- Custom filters allow us to create filters that the designer did not supply us with. We create filters with objects and
  - the different logical operators that are available.
  - the different types of operands in BusinessObjects.

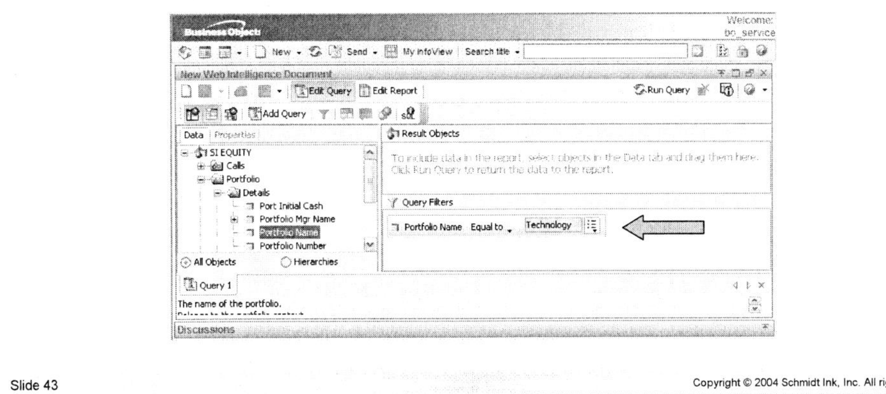

Slide 43

The Designers can create many query filters, but they probably can not anticipate every need for a filter in our documents. Therefore, Web Intelligence allows us to create our own custom filters.

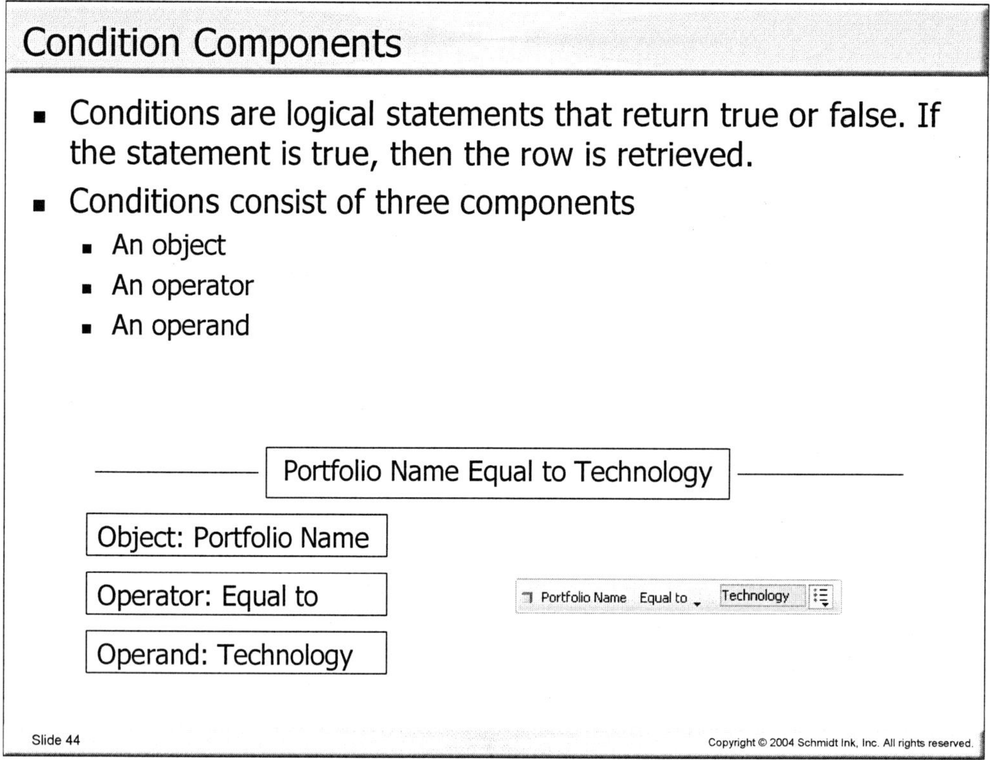

Condition Components
Conditions are logical statements that return true or false. If the statement is true, then the row is retrieved.
Conditions consist of three components
An object
An operator
An operand
Portfolio Name Equal to Technology
Object: Portfolio Name
Operator: Equal to
Operand: Technology
Portfolio Name Equal to Technology
Slide 44
Copyright © 2004 Schmidt Ink, Inc. All rights reserved.

Most conditions contain three different components. An object, an operator, and an operand.

Creating a Custom Query Filter

- Select the object for a Query Filter by dragging it from the Data tab into the Query Filters window. Objects may also be dragged from the Result Objects section.
  - Almost any object can be used in a filter. Although, the universe designer may not allow certain objects, because they may slow query performance too drastically.

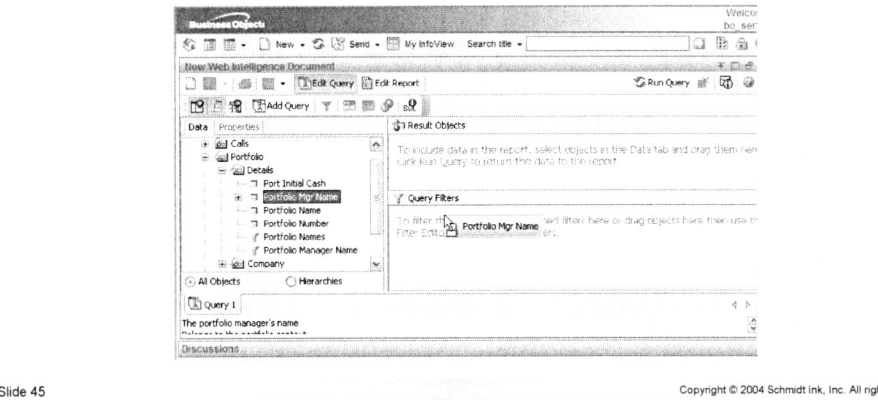

Slide 45

To select an object for a filter, we simply drag it from the Data tab and drop it into the Query Filters window. You may also drag it from the Result Objects window.

Creating a Custom Query Filter

1. Create a new query with Portfolio Name, Portfolio Mgr Name, Port Initial Cash
2. Drag the Portfolio Name object into the Query Filters section.
3. Continue on next slide…

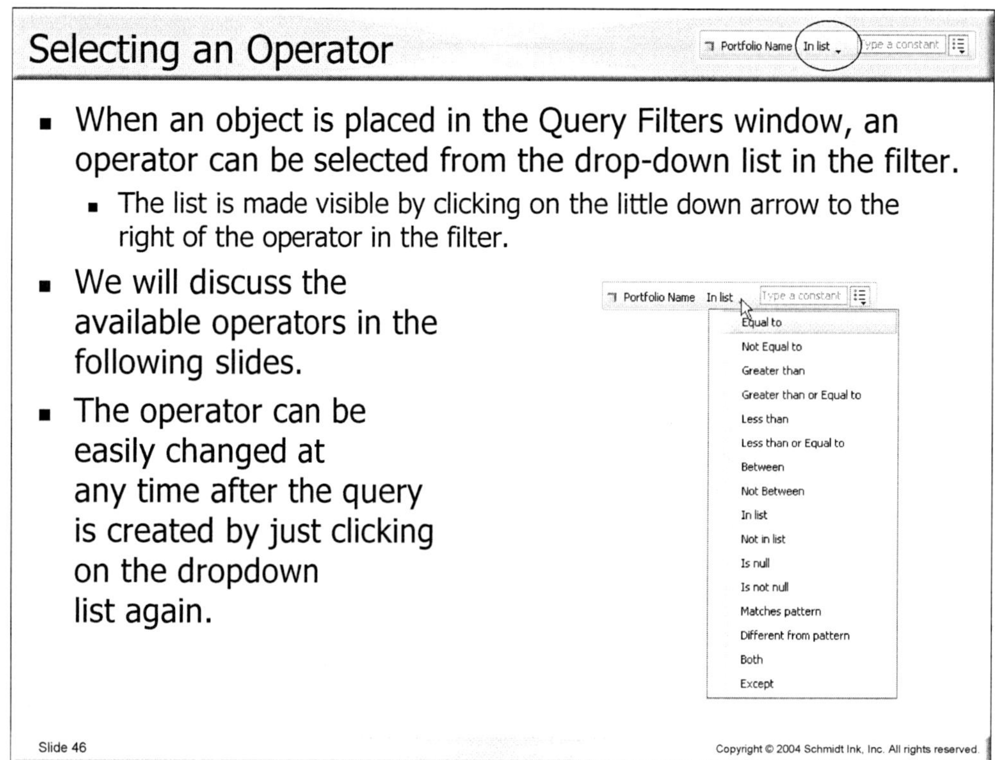
Selecting an Operator
Portfolio Name In list Type a constant
When an object is placed in the Query Filters window, an operator can be selected from the drop-down list in the filter.
The list is made visible by clicking on the little down arrow to the right of the operator in the filter.
We will discuss the available operators in the following slides.
The operator can be easily changed at any time after the query is created by just clicking on the dropdown list again.
Portfolio Name In list Type a constant
Equal to
Not Equal to
Greater than
Greater than or Equal to
Less than
Less than or Equal to
Between
Not Between
In list
Not in list
Is null
Is not null
Matches pattern
Different from pattern
Both
Except
Slide 46
Copyright © 2004 Schmidt Ink, Inc. All rights reserved.

The available operators allow to create Query Filters that allow us to bring back data of interest from the database.

1. Select the *Equal to* operator.
2. Continue on next page…

Selecting an Operand Type

- BusinessObjects offers the following types of operands
  - Constant
    A constant value is typed in to the Query Filter. This value does not change, unless the query is edited and another value is typed in.
  - Value(s) from a list
    This operand is also a constant value, but this value is selected from a list of available values. This has the advantage of always being a value that exists in the database.
  - Prompt
    Allows the viewer of the report to select or enter a value when the query is refreshed. This allows the viewer to enter values without editing the query filter.
  - Object
    This operand assigns a value of another object to the operand. For this operand to work, the object must be logically related to the object of the Query Filter.

Slide 47

We can select an operand type from the drop-list of operands in the Query Filter. The default operand is the constant operand and we can just type a value into the operand portion of the Query Filter.

The risk of using constant operands is that the value may not exist in the database. Also, some databases are case-sensitive. This means that *robert ≠ Robert*, because the first letters are of differing cases.

1. Click on the operand portion of the Query Filter to activate the operand edit field..
2. Type Technology in the edit field on the object.

We have now created a simple condition on the Portfolio Name object.

Operator - Equal to / Different From

- The Equal to operator allows information to be returned when the object portion of the condition exactly matches the operand portion.
- In many systems this match includes the case of the word.
  - For example, CASE = CASE and CASE ≠ Case.
- The Different From operator allows information to be returned when the object portion of the condition does not exactly match the operand portion.
  - This is the same as Not Equal To.

Slide 48

The 'Equals to' operator returns rows when an object equals a certain value. This is the type of logic that we use most often. However, sometimes we are interested in retrieving rows when an object is not equal to a certain value. For example, *Revenue/Expense Different from 0*.

The different from operator allows us to eliminate unwanted information from our query. For example, suppose that a developer put a test account in our production warehouse. It would be quite annoying if this bogus account appeared in our corporate documents. If this is the case, we could place *Account Different from Bogus Account* and ignore this account in the universe.

Operator – Range Operators

- The range operators allow us to return information when the object value is greater than, less than or even between the operand value(s).
- Range operators work with numbers, dates or text. Below are some examples
  - Number Greater than 5
    Will return 6, 7, 8, ...
  - Number less than or equal to 5
    Will return ..., -1, 0, 1, 2, 3, 4, 5.
  - Name Greater than 'B'
    Will return Bertha, Betty, ...
  - Name Between 'B' and 'C'
    Will return all names that start with B
  - Date Between 1/1/99 and 1/31/99
    Will return all data for January 1999

Slide 49

Range operators allow us to bring back information when a range of conditions are true. For example, *Revenue/Expense Greater than 100000.*

As shown above, the range operators also work on text values. Remember in college during registration? At my school they would divide the people up into groups defined by their last names.

Name Greater than 'A' And Name Less than 'F'

Or

Name Greater than or Equal to 'F' And Name Less Than 'R'

Or

Name Greater than or Equal to 'R'

Operator - Matches Pattern

- The Matches Pattern operator allows users to specify wild cards in their conditions.
- For example, if a report developer wanted to return all customers whose name started with a B, the developer could use the following
  - Customer Matches Pattern 'B%'
  - The percent sign represents any number of unknown characters.
- Suppose a developer wanted to know three letter names that start with a B and end with an R. The developer could then use the following
  - Customer Matches Pattern 'B_r'
  - The underscore represents one unknown character.

Slide 50

The Matches Pattern operator is a great operator that allows us to define wild cards for text conditions. For example, suppose we wanted all Companies that started with the letter A. We could build a condition like, Equity Price Company Matches Pattern 'A%'.

This example is relatively efficient and it will return all companies that start with the letter 'A'. Now suppose that we wanted to create a condition that returned all products that end with the pattern – '5AH'? We could do the following: Product Number Matches Pattern '%5AH'. This query is not very efficient, because it can not use an index to locate the values in the database.

An analogy to the above examples would be if I asked you the show me all of the words in a dictionary that started with the letter 'A'. Easy enough, you just show me those pages. However, if I asked you to show me all of the words that ended with 'ion', it would be near impossible. If you requested this information from a database, it would take much longer than simply searching for words that start with the letter 'A'.

Operator – Is Null / Is Not Null

- A Null is an unassigned value. This missing assignment can be explicit or implicit.
  - Explicit: The value was never assigned in the database. For example, in a contact management system you may leave the cell phone field empty.
  - Implicit: Supposed you chose the Customer and Invoice objects from a sales universe. The Customer object represents all of your customers and the invoice object represents all of your invoices, both are completely assigned. However, when you select both together, there may be customers with no invoices. These customers will have Null invoice numbers.
- The Is Null operator allows you to retrieve rows where an object's value is Null.
- The Is Not Null operator allows you to return rows where an object's value is not Null.

Slide 51

This is a very important operator, as it allows you to retrieve or ignore rows that have Null values. Null values are fields that have never been assigned a value.

Operand – Constant

- To enter a constant value for an operand, just click on the operand portion of the Query Filter, then type in a value.
  - This value must exactly match the value that is in the database. Below are some reasons the values may not match
    - Values have different case Robert ≠ robert (not all data sources are case sensitive.)
    - The value in the database may have spaces appended to it 'Robert' ≠ 'Robert '. This is actually quite common.
    - Dates are stored as text, not dates. 5/5/2005 ≠ '5/5/2005'. This is also quite common.
    - The value may simply be misspelled Robert ≠ Robret. This is very common and a source of frustration, because it takes a while to realize that the value has been misspelled.

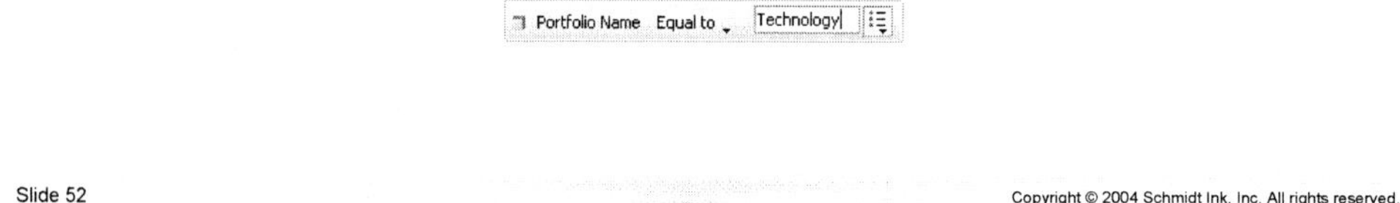

Slide 52

Now that we know all of the available operators, it is time to discuss the operands.

The first of the operands, and maybe the most common, is *Constant*. This operand allows you to type in the value for the operand. It is quick and easy, but sometimes also error prone. It is error prone, because you may mistype the value. For example, suppose the condition was on quarters in a year and you entered Q4. Seems reasonable, but maybe the quarters are not stored that way. Maybe they are stored like Qtr 4. These two values are not the same and no data would be retrieved.

Operand – Value(s) from List

- *Value(s) from list* also allows us to enter a constant value for the operand. However, instead of typing the value, we are allowed to select it from a list.
  - Selecting values ensures that they will be of the same case and spelled the same. Therefore, eliminating human error. It also ensures that if there are trailing spaces, they will also be appended to the operand value, thus ensuring a match.
  - The list may take significantly longer to display the first time it is used, because Web Intelligence must run a query to retrieve the values.

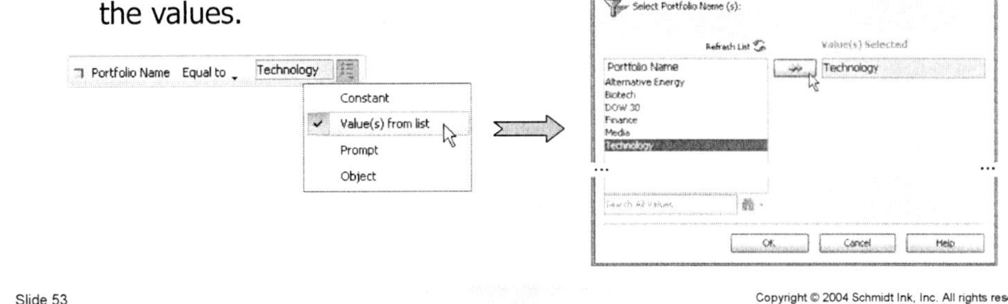

Slide 53

The *Value(s) from list* operand is very similar to the type a new constant operand, as both place constants in the operand. However, the *Value(s) from list* operand allows you to select a value or values from a list.

Web Intelligence must run a query to populate the list the first time it is displayed for an object. In some cases, this could take a relatively long time, and in others not so bad. The good news is that once the list is created, it may not have to update, until you click the Refresh List button in the List of Values dialog.

If a value is missing from the list, but you know that it has recently been added to the database, such as a new customer, a new region, a new product, and so forth, then you can click the Refresh List button. This button refreshes only the values in the list and has no effect on the data in the report.

Not all objects have a list of values associated with them. For example, most measures do not have a list of values. Neither do large text fields.

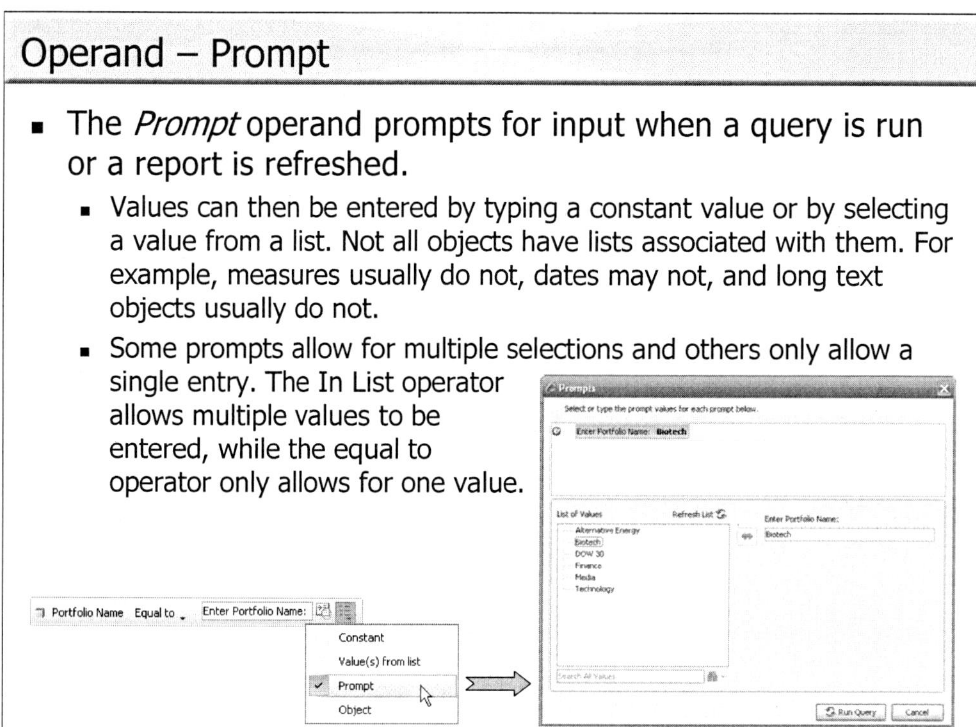

Operand – Prompt
The *Prompt* operand prompts for input when a query is run or a report is refreshed.
Values can then be entered by typing a constant value or by selecting a value from a list. Not all objects have lists associated with them. For example, measures usually do not, dates may not, and long text objects usually do not.
Some prompts allow for multiple selections and others only allow a single entry. The In List operator allows multiple values to be entered, while the equal to operator only allows for one value.
Portfolio Name
Equal to
Enter Portfolio Name:
Constant
Value(s) from list
Prompt
Object
Prompts
Enter Portfolio Name: Biotech
List of Values
Refresh List
Alternative Energy
Biotech
DOW 30
Finance
Media
Technology
Enter Portfolio Name:
Biotech
Run Query
Cancel
Slide 54
Copyright © 2004 Schmidt Ink, Inc. All rights reserved.

The prompt is an extremely flexible way to allow people to enter values for a condition, because people do not have to edit the query to change the operand value. When a document is refreshed, the prompts dialog is displayed and then people can just accept the defaults or select new values. They can type a new value, or in most cases, select a value from a list.

Operand – Object

- BusinessObjects allows other objects to be used as an operand. This allows the condition to equate the values of two different objects.
  - The outcome of this type of operand is often unpredictable, because it creates a relation with another object in the universe. Knowledge of the tables in the universe and their relations will help to eliminate unpredictable behavior.
  - Any time the operand is another object, the result set must be analyzed to make sure it is returning the anticipated data.

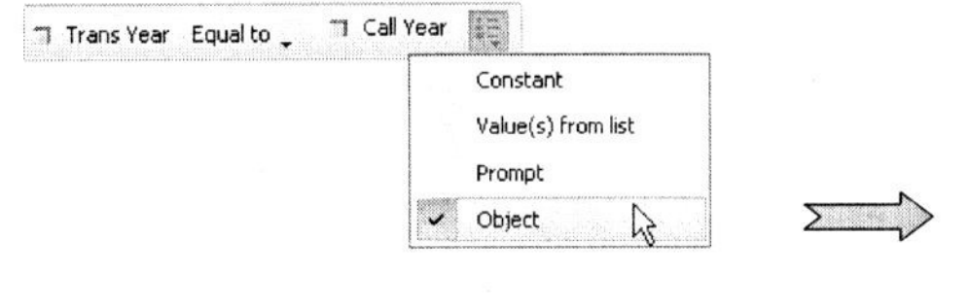

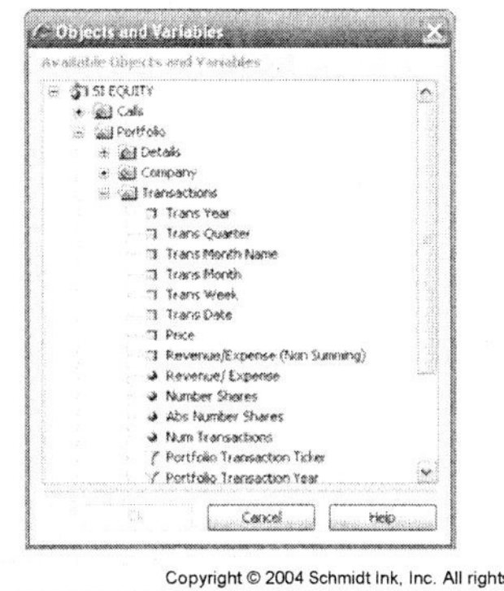

Slide 55

We can use other objects as an operand. Sometimes this is a good idea, sometimes it is not. The reasons why are too technical for the scope of this course, but it is important to realize, so that you always check the validity of the results returned when an object is used as an operand in a condition.

Okay, I will tell you why. Those of you that know SQL know that we join tables in a query by equating two fields in two different tables. For example, Trans Year equal to Call Year. In most cases, the universe designer has carefully considered all joins in the universe and made all of the joins that will allow data to be safely retrieved. Now, when you create a condition equating one object to another, you just created a new join in the universe. Is it right? Could be....

When you become more advanced, you will learn to use subqueries. In most cases, you could use a subquery to get the results that you wanted to get by using an object as an operand.

- Add a Subquery

are queries that supply a list of values for an operand n. In Web Intelligence we use an object in the Data this list.

suppose that we wanted to use Trans Date to supply a for a Call Date condition. This would be similar to Where in list Trans Date.

his may seem similar to using an object in the operand, but since the Trans Date is in a subquery, it is independent of the objects in the Result Objects section.

- To create a subquery, we select an object by clicking on it, and then we click the Add a Subquery button. This will place the subquery framework in the Query Filters section.

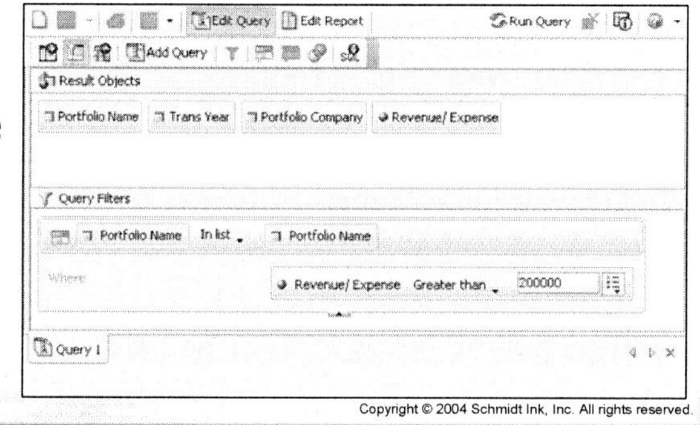

Slide 56

Subqueries allow us to create queries that are independent of the resolution or context of the condition. For example, we can create a subquery that is more summarized than the combination of the objects in the Result Objects window. We will do this example over the next few slides. We can also create conditions that allow us to use objects from different contexts of our universe. We will do this example following the summary condition example.

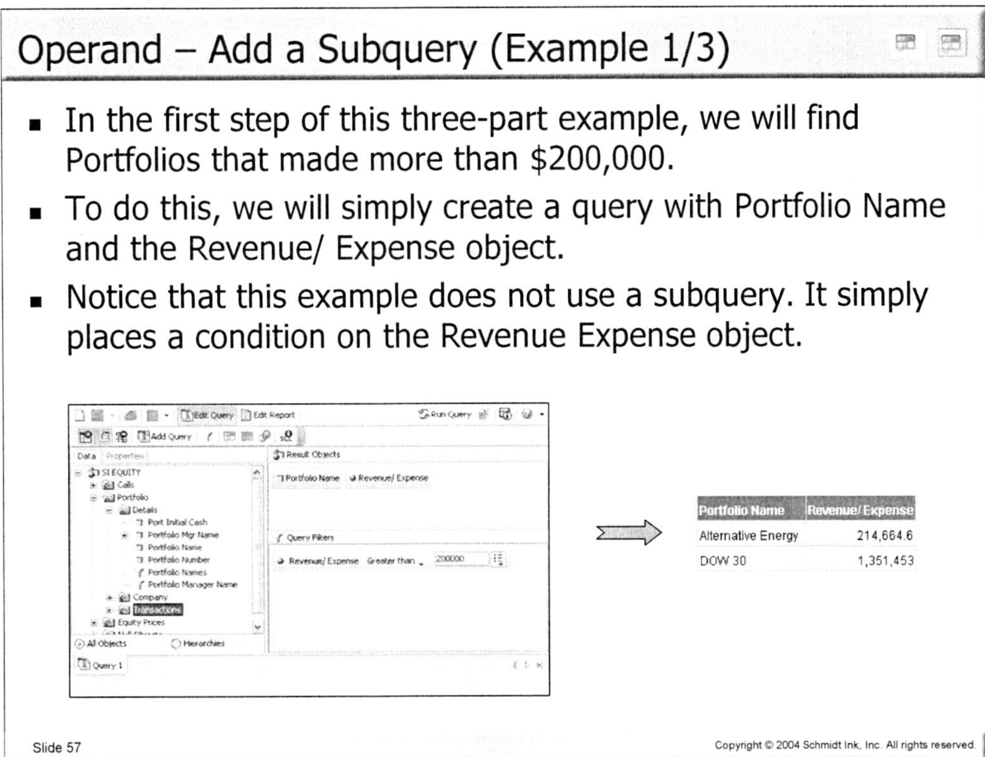

This first example retrieves Portfolio Names that generated more than $200,000. It is a simple query. We are going to use this example to prove that we need subqueries to get more detailed information than just a portfolio name. In addition, we will see that a query similar to this query will be used as a subquery to help retrieve the detailed information.

Create a Query Displaying Portfolios that Made More Than $200,000

1) Select New>Web Intelligence Document from the menu.
2) Select the SI Equity universe.
3) Double-click Portfolio Name and Revenue/ Expense in the Portfolio Class, to place them in the Result Objects pane.
4) Drag Revenue/ Expense into the Query Filters pane.
5) Select the *Greater than* operator.
6) Type 200000 in the operand portion of the condition.

Operand – Add a Subquery (Example 2/3)

- In the second part of this example, let's find Portfolios that made more than $200,000 and the amounts they made from each company in the portfolio.
- To do this, we will create a new query with Trans Year and Portfolio added to the previous example.
    - In the previous example, we found out that two portfolios made more than $200,000. However, in this example, the result set only contains one portfolio.

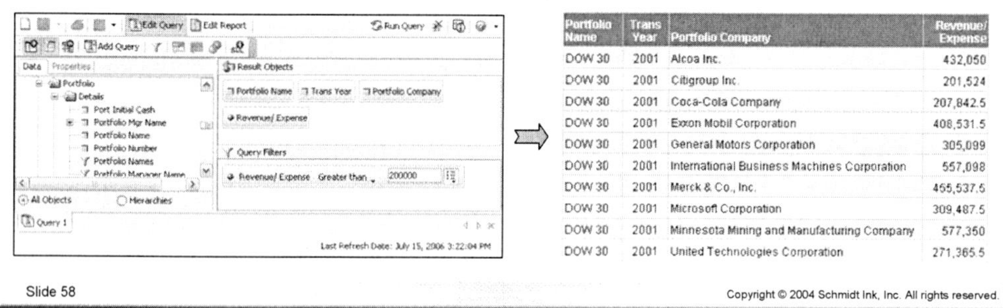

Portfolio Name	Trans Year	Portfolio Company	Revenue/ Expense
DOW 30	2001	Alcoa Inc.	432,050
DOW 30	2001	Citigroup Inc.	201,524
DOW 30	2001	Coca-Cola Company	207,842.5
DOW 30	2001	Exxon Mobil Corporation	408,531.5
DOW 30	2001	General Motors Corporation	305,099
DOW 30	2001	International Business Machines Corporation	557,098
DOW 30	2001	Merck & Co., Inc.	455,537.5
DOW 30	2001	Microsoft Corporation	309,487.5
DOW 30	2001	Minnesota Mining and Manufacturing Company	577,350
DOW 30	2001	United Technologies Corporation	271,365.5

Slide 58

This example returns an erroneous data set. We thought that we were creating a query that will return the portfolios that made more than $200,000 and some of the details of the portfolio transactions. However, this is not what the query returned. The query returned Portfolio Companies that made more than $200,000 in a transaction year. This query is fundamentally different from the one in the previous example, because the extra dimension objects change the structure of the query. They cause the Query Filter to be applied to more detailed objects than in the previous example. In the next slide, we will solve this problem with a subquery.

Create a Query Displaying Portfolios that Made More Than $200,000 (Erroneous)

1) Select New>Web Intelligence Document from the menu.
2) Select the SI Equity universe.
3) Double-click Portfolio Name, Trans Year, Portfolio Company and Revenue/ Expense in the Portfolio Class, to place them in the Result Objects pane.
4) Drag Revenue/ Expense into the Query Filters pane.
5) Select the *Greater than* operator.
6) Type 200000 in the operand portion of the condition.

Operand – Add a Subquery (Example 3/3)

- In the previous example, we found out that we could not retrieve the information that we were interested in, by using a simple query. In this example, we will solve the problem by adding a subquery to the Query Filters section.
- This solution works because it creates a list independently of the objects in the Result Objects window. Therefore, the subquery will create a list of portfolios that did generate more than $200,000.
  - This type of query is known as retrieving detail information from a summary condition.

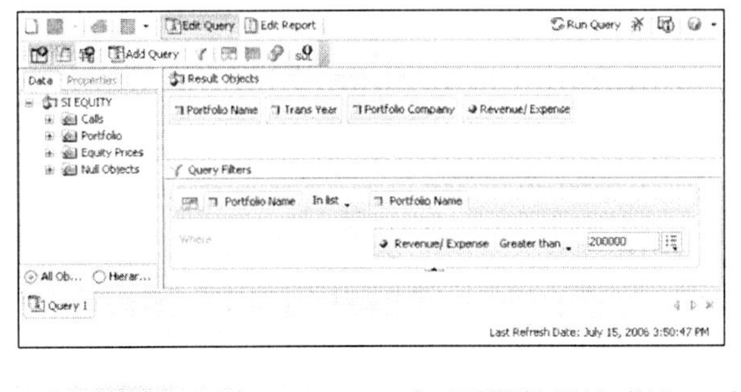

Slide 59

The first of these three examples allowed us to determine which portfolios generated more than $200,000. In the second example, we added more dimension objects, because we wanted more details in our report. However, when these extra details were added, the revenue measure object recalculated to find the totals for each combination of Portfolio, Year, and Company. Then, the condition was applied to these new totals, rather than just the portfolio totals. To correct this behavior, we had to make a condition that only looked at portfolio totals. To do this, we used a subquery. This subquery returned a list of companies that generated more than $200,000.

Create a Query Displaying Portfolios that Made More Than $200,000

1) Select New>Web Intelligence Document from the menu.
2) Select the SI Equity universe.
3) Double-click Portfolio Name, Trans Year, Portfolio Company and Revenue/ Expense in the Portfolio Class, to place them in the Result Objects pane.
4) Click the *Add a Subquery* button.
5) Drag the Portfolio Name Dimension object into the two empty fields on the subquery object. The subquery should read: Portfolio Name In List Portfolio Name.
6) Drag Revenue/ Expense onto the condition portion of the subquery object in the Query Filters pane.
7) Select the *Greater Than* operator.
8) Type 200000 in the operand portion of the condition.

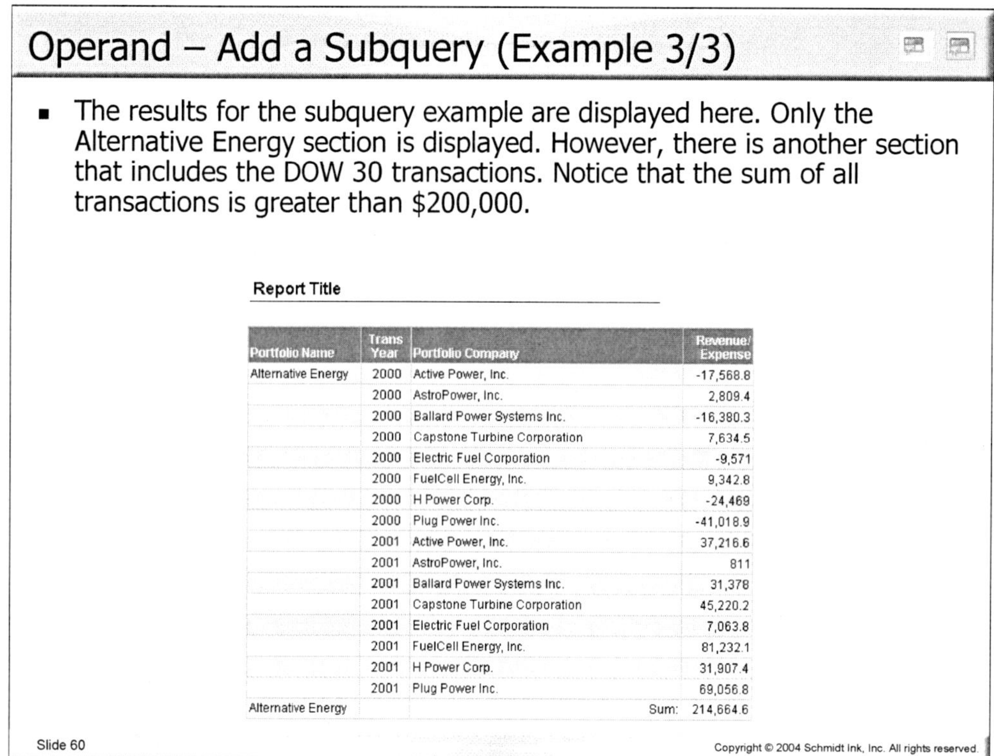
Operand – Add a Subquery (Example 3/3)
The results for the subquery example are displayed here. Only the Alternative Energy section is displayed. However, there is another section that includes the DOW 30 transactions. Notice that the sum of all transactions is greater than $200,000.
Report Title
Slide 60
Copyright © 2004 Schmidt Ink, Inc. All rights reserved.

Portfolio Name	Trans Year	Portfolio Company	Revenue/ Expense
Alternative Energy	2000	Active Power, Inc.	-17,568.8
	2000	AstroPower, Inc.	2,809.4
	2000	Ballard Power Systems Inc.	-16,380.3
	2000	Capstone Turbine Corporation	7,634.5
	2000	Electric Fuel Corporation	-9,571
	2000	FuelCell Energy, Inc.	9,342.8
	2000	H Power Corp.	-24,469
	2000	Plug Power Inc.	-41,018.9
	2001	Active Power, Inc.	37,216.6
	2001	AstroPower, Inc.	811
	2001	Ballard Power Systems Inc.	31,378
	2001	Capstone Turbine Corporation	45,220.2
	2001	Electric Fuel Corporation	7,063.8
	2001	FuelCell Energy, Inc.	81,232.1
	2001	H Power Corp.	31,907.4
	2001	Plug Power Inc.	69,056.8
Alternative Energy		Sum:	214,664.6

The above report has been formatted to better show the results. We will learn how to format reports later in this course.

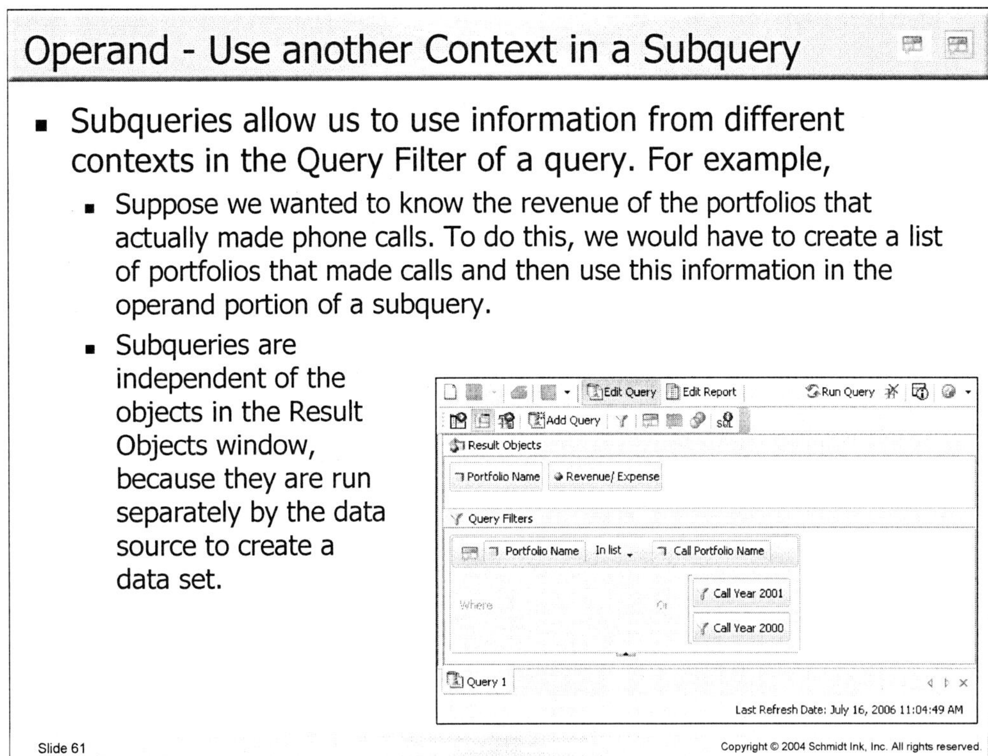

Operand - Use another Context in a Subquery
Subqueries allow us to use information from different contexts in the Query Filter of a query. For example,
Suppose we wanted to know the revenue of the portfolios that actually made phone calls. To do this, we would have to create a list of portfolios that made calls and then use this information in the operand portion of a subquery.
Subqueries are independent of the objects in the Result Objects window, because they are run separately by the data source to create a data set.
Edit Query
Edit Report
Run Query
Add Query
Result Objects
Portfolio Name
Revenue/ Expense
Query Filters
Portfolio Name
In list
Call Portfolio Name
Call Year 2001
Call Year 2000
Query 1
Last Refresh Date: July 16, 2006 11:04:49 AM
Slide 61
Copyright © 2004 Schmidt Ink, Inc. All rights reserved.

Subqueries are ran independently of the rest of the query. They are used to create a data set that can contain one or many values. Once the set is created, the Query Filter will then use the set in the operand portion of a condition. In this case, it will look for portfolio names that are in the list.

Create a List of Portfolios (and their Revenues) that Made Phone Calls

1) Select New>Web Intelligence Document from the menu.
2) Select the SI Equity universe.
3) Double-click Portfolio Name and Revenue/ Expense in the Portfolio Class, to place them in the Result Objects pane.
4) Click the *Add a Subquery* button.
5) Drag the Portfolio Name Dimension object into the first empty field (the leftmost) on the subquery object.
6) Open the Calls Class and find the Call Portfolio Name dimension object in the Multiple class.
7) Drag the Call Portfolio Name dimension object to the second field (The rightmost).
8) Find the Call Year 2001 and Call Year 2000 Query Filter objects and drop them onto the Filter portion of the subquery. Make sure that an *Or* is used as a conjunction operator. If it is an *And*, then simply double-click on it to turn it into an *Or*.

   We need to use these filters to force BusinessObjects to only return only Call Portfolio Names that have made calls. Without the filters, all of the Call Portfolio Names will be returned.

9) Click the Run Query button to find out how much revenue the portfolios that made phone calls generated.

Query Filters Summary

- In this chapter
  - We learned about Designer defined Query Filters.
  - We learned how to combine multiple conditions in a query.
  - We learned how to create our own custom Query Filters.
  - We examined the components of Query Filters
    - Object, Operator, and Operand.
  - We talked about the different Operators and Operand types.
  - We explored a couple queries that use subqueries
    - Detail information from a summary condition
    - Using a list created from a different context.

Slide 62

In this chapter, we have learned that query filters are very important, because they allow us to target information in our reports. This makes our reports more concise and more readable. These qualities make our reports very valuable to our company and the employees that use them in their business making decisions.

Creating Documents with Web Intelligence XI

Working with Report Structures
(Java Report Panel)

Introduction

- Now that we know how to retrieve data for our documents, it is a good time to learn how to format the data into useful information. One of the first lessons that a good report developer learns is that data must be formatted before it becomes useful to the company.
- In this chapter we are going to learn how to create tables, crosstabs, charts, and other report structures. We are also going to learn how to modify existing reports and structures to create more powerful report presentations.

Slide 64

Good reports contain information that is useful to a company. Great reports present the information in a way that is readable and easy to comprehend. Employees appreciate reports that allow them to easily proceed in their business making decisions. In this chapter we are going to learn how to present information in Web Intelligence reports.

Data Types (Dimension)

- Earlier in the Query chapter, we learned about the different data types in Web Intelligence. We will now discuss the behavior of each type.
  - Dimension
    - Uniquely identifies nouns in a report. Dimensions are important, because they define the context of a row. Contexts define the framework for the calculations in a report.
    - For example, suppose we have table with only Employee and Total Compensation. The context of each row is Employee. The Total Compensation conforms to display the total for each Employee.
    - Now suppose the table has Department, Employee and Total Compensation. The context of each row is now Department and Employee.
      - One could argue that Employee defines the context, but what about the cases where an Employee may be compensated by different Departments?

Slide 65

Dimensions define the contexts in our reports. The are the nouns of interest. For example, we may have the Regional report - Region is a dimension. We may have the Region, Salesperson, Product report – Region, Salesperson, and Product are dimensions. The dimensions form the framework for the report.

Data Types (Detail) 

- Details support the dimensions in a report. There is usually only one detail value for each dimension value. In fact, Business Objects assumes this to be true, although many universe designers violate this assumption.
  - For example
    - Each car can only be one color at a time.
    - An employee can only have one name.
    - An employee can only have one desk phone number.
  - If this rule is adhered to, then we can use this assumption to create more flexible reports as we become more advanced.
  - Details do not create contexts, since there is only one value for each dimension value.
    - This is why we do not Merge Details, as we did Dimensions in the Query chapter.

Slide 66

Details are useful in reports, because they do not define contexts. This means that they do not have the same restrictions that dimensions do. At our current level, this may not seem to be important, but as we become more advanced, we will appreciate this fact.

Many people do not understand the significance of details and their behavior within reports. In fact, many universes do not even have a detail object. Will this stop us from creating our reports? No, but it can make some reports more difficult to create. I like to think of a detail, as something like a spy. It can be placed in tables and have no effect on the context of the table. Therefore, Web Intelligence may let a detail into tables where it will not let an unmerged dimension.

Data Types (Measure)
- Measures perform most of the aggregate calculations in a report. They will always (mostly) roll up to the context defined by the dimensions in a row.
  - The salary object used in the Dimension slide is probably a measure.
    - If we have a report with Employee and Salary, then the Salary measure will roll up and display the salary for each employee.
    - If we have a report with Territory and Salary, then that same Salary measure will conform to the Territory and display the total salary in each Territory.
    - In both of the above examples, Employee and Territory define the contexts of their respective tables. The Salary measure simply conforms to the context and displays the proper total.
  - Most measures have an aggregate function associated with them. This function allows them to conform to the contexts on a report. For example, they can Sum, Count, Max, Min, Average, and so forth.
Slide 67
Copyright © 2004 Schmidt Ink, Inc. All rights reserved.

Measures will usually conform to the context in which they are placed. They can sum, count, min, max, and average. Most reports contain both measures and dimensions. Details are often overlooked and treated as dimensions. In most cases, this is okay. However, in a few cases this will make reporting more difficult.

Knowing how measures behave within contexts will allow us to create very powerful reports as we become more advanced report writers. It will also allow us to quickly create reports without a lot of trial and error.

Report Formats and Structures

- Now that we know about the different data types that objects can be, we are going to use these objects to create tables, crosstabs, charts, and freestanding cells.
- Each of these types or structures has their advantages in the reporting world and no one report type can cover all reporting needs. Many people worry that they will not know what type to use, but with experience the choice becomes easier to make.

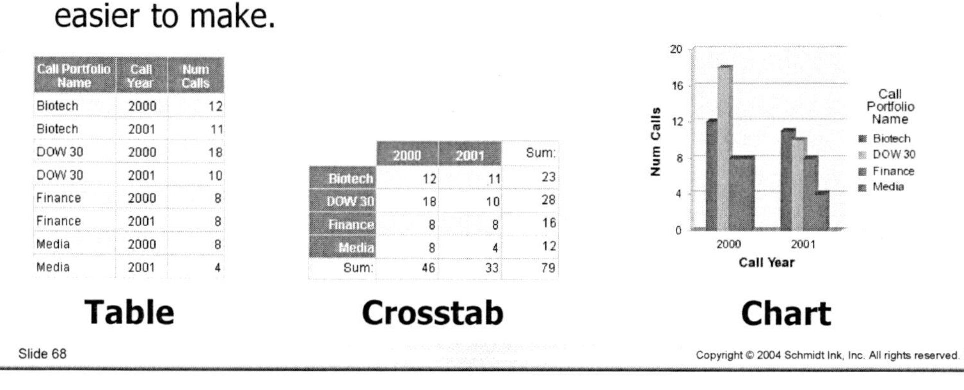

Call Portfolio Name	Call Year	Num Calls
Biotech	2000	12
Biotech	2001	11
DOW 30	2000	18
DOW 30	2001	10
Finance	2000	8
Finance	2001	8
Media	2000	8
Media	2001	4

	2000	2001	Sum:
Biotech	12	11	23
DOW 30	18	10	28
Finance	8	8	16
Media	8	4	12
Sum:	46	33	79

Slide 68

Web Intelligence gives us several report structures to use in our reports. Each type is very useful and each has its own advantages and disadvantages. As you become more advanced, you will use each of the report types to create more and more powerful reports. A powerful report is one that allows viewers of the report to understand exactly what you are trying to show them. The best reports are concise and clear. They show the viewers of the document how to improve their business. We should always create reports with the viewers in mind.

Tables

- Tables consist of rows of information
  - Each row is defined by a unique combination of the dimension and detail values, which means that there should be no two rows in a table that are identical
  - If a table contains a measure, it will conform to the rows of the table.
    - Each measure usually has a default aggregate function assigned to it. This causes them to always sum, average, min, max, or count to conform to the dimension defined rows.
- Most tables have a header row
  - Header rows are not defined by combinations of dimensions
  - Header rows are singular in nature
  - Header rows typically contain labels for the columns of the table

Portfolio Name	Trans Year	Revenue/Expense
Alternative Energy	2000	(89,221)
Alternative Energy	2001	303,886
Biotech	2000	(369,888)
Biotech	2001	449,314
DOW 30	2000	(4,385,413)
DOW 30	2001	5,736,866
Finance	2000	(507,101)
Finance	2001	574,563
Media	2000	(465,394)
Media	2001	386,854
Technology	2000	(1,116,767)
Technology	2001	1,059,497

Slide 69

Almost all reports start out as a table. Because it is the default report structure. This makes sense, because a table can easily be converted into any of the other report structures.

In the above slide, we stated that rows are defined by Dimensions and Details. While this can be true, in most cases rows are defined by the Dimensions in the row, because there should only be one detail value for each dimension value. In some universes, the designer has called an object that has multiple values per dimension value a detail. In these cases, the detail must be treated as a dimension.

Inserting Tables

- Whenever we create a new report, the default structure is a table (in most installations). Sometimes a single table is enough for a report and sometimes we need several tables.
- We can add tables to a report by inserting a new table, or by copying an existing table.
  - To insert a table, we drag multiple objects from the Manager onto a report.
  - Later, in this chapter, we will learn how to copy an existing table.

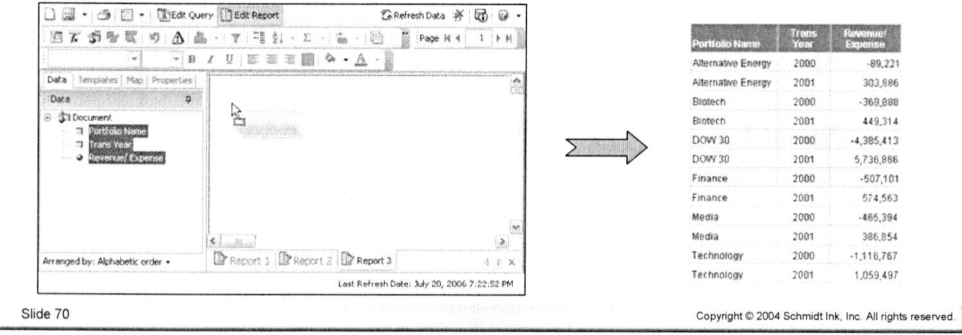

Portfolio Name	Trans Year	Revenue/ Expense
Alternative Energy	2000	-89,221
Alternative Energy	2001	303,886
Biotech	2000	-369,888
Biotech	2001	449,314
DOW 30	2000	-4,385,413
DOW 30	2001	5,736,886
Finance	2000	-507,101
Finance	2001	574,563
Media	2000	-465,394
Media	2001	386,854
Technology	2000	-1,116,767
Technology	2001	1,059,497

Slide 70

Almost all reports start out as a table. Once created, tables can be transformed into charts, crosstabs, and rotated tables. We can also create new tables by dragging objects from the Data tab of the Manager and dropping them on a report. However, if you only drag a single dimension object into a report that already has a table, then it will not create a single column table. Dragging a single dimension object onto a report with a table will create sections on the report. We will learn more about sections later in this course.

Create a Table on a Report

1. Create a new report with Portfolio Name, Trans Year, Revenue/ Expense, and Num Transactions
   Notice that the default report structure is a table
2. Insert a new report, by right-clicking on the report tab and choosing Insert Report from the pop-up menu
3. Select Portfolio Name, Trans Year and Revenue/ Expense from the Data tab of the Manager
4. Drag the selected objects onto the report
   The cursor will show that you are dragging a cluster of objects. This cursor tells you that a table will be created.
5. The newly dropped items will create a new table.

Selecting a Table

- To select a table, move the cursor to just outside any border of the table. The table will highlight with a grey halo. Click on the halo to select the table.
  - It is important to select a table in this manner, because to format a table, the entire table must be selected, not just a cell or column.
  - Notice that when a table or report structure is selected, a border appears around the structure and no individual cells in the table are selected.

Portfolio Name	Trans Year	Revenue/Expense
Alternative Energy	2000	-89,221.3
Alternative Energy	2001	303,885.9
Biotech	2000	-369,888.4
Biotech	2001	449,314.1
DOW 30	2000	-4,385,412.5
DOW 30	2001	5,736,865.5
Finance	2000	-507,100.7
Finance	2001	574,563
Media	2000	-465,393.9
Media	2001	386,853.6
Technology	2000	-1,116,767.25
Technology	2001	1,059,497.45

Slide 71

When creating reports in Web Intelligence, we often have to select tables. We select tables to copy or move them, to assign property values to them, to transform them into other report structures, and so forth.

1) Select the table in the report from the previous exercise.

Sizing Rows and Columns

↔

- Columns and Rows in a table can be sized by placing the cursor over the border, and then when the cursor changes to the Size Cursor (↔), we click and drag the border to a new position.
- We almost always have to adjust the size of the columns and rows in a report to conform the cells to the size of the data that they contain.

Portfolio Name	Trans Year
Alternative Energy	2000
Alternative Energy	2001
Biotech	2000
Biotech ↔	2001
DOW 30	2000
DOW 30	2001

Adjusting the column width.

Portfolio Name	Trans Year
Alternative Energy	2000
Alternative Energy	2001
Biotech	2000
Biotech	2001
DOW 30	↕ 2000
DOW 30	2001

Adjusting the row height.

Slide 72

We almost always have to adjust the column widths and row heights in a report. If we do not, the report may not look professional. In the business of creating reports, we must always strive to create reports that are professionally formatted. Reports that are formatted with care, help give the viewers confidence in the report.

Removing Columns

- To remove a column from a table
  - Right-click on the column and select Remove Column from the pop-up menu.
  - If you delete a dimension column, the measures in the table will conform to the remaining dimensions in the table and properly recalculate.

Portfolio Name	Trans Year	Revenue/ Expense	Num Transactions
Alternative Energy	2000		
Alternative Energy	2001		
Biotech	2000		
Biotech	2001		
DOW 30	2000		
DOW 30	2001		
Finance	2000		
Finance	2001		
Media	2000		
Media	2001		
Technology	2000		
Technology	2001		

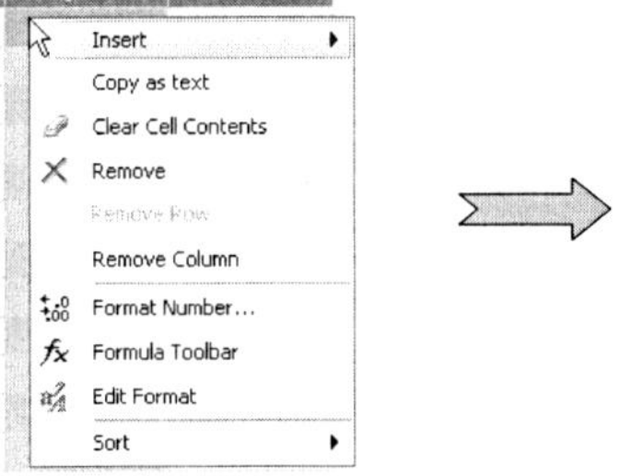

Insert
Copy as text
Clear Cell Contents
Remove
Remove Row
Remove Column
Format Number...
Formula Toolbar
Edit Format
Sort

Portfolio Name	Trans Year	Num Transactions
Alternative Energy	2000	78
Alternative Energy	2001	96
Biotech	2000	91
Biotech	2001	93
DOW 30	2000	675
DOW 30	2001	816
Finance	2000	45
Finance	2001	75
Media	2000	62
Media	2001	58
Technology	2000	274
Technology	2001	317

Slide 73

We often remove columns from tables. We do this to simply a table before converting it to another report structure, to remove un-needed objects, or various other reasons.

Remove a Column

1) Create a report with Portfolio Name, Trans Year, Revenue/Expense, and Num Transactions.
2) Right-click on the Trans Year column and select *Remove column* from the pop-up menu.
   Notice that the Revenue and Num Transactions measure objects conform to the new context defined by Portfolio Name.

Inserting Columns

- To insert a column
  - Click on a column in the table to select it
  - Click on Insert Column... in the Reporting toolbar.
  - You can also right-click on a column and chose Insert... from the pop-up menu.
- We often insert columns into reports to add calculations and other informational formulas.

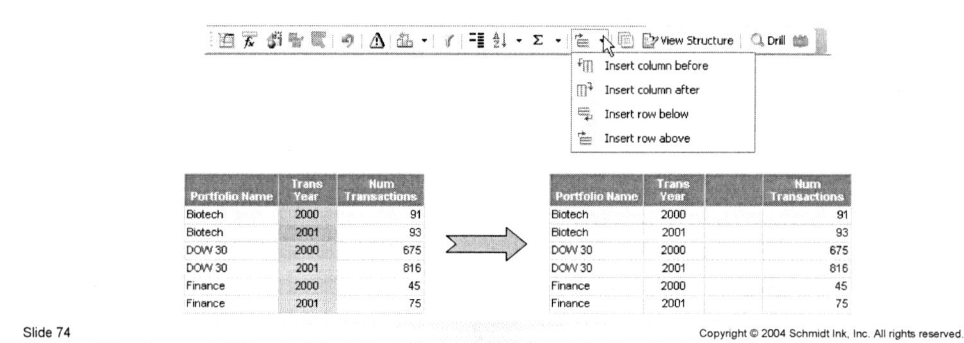

Portfolio Name	Trans Year	Num Transactions
Biotech	2000	91
Biotech	2001	93
DOW 30	2000	675
DOW 30	2001	816
Finance	2000	45
Finance	2001	75

Portfolio Name	Trans Year		Num Transactions
Biotech	2000		91
Biotech	2001		93
DOW 30	2000		675
DOW 30	2001		816
Finance	2000		45
Finance	2001		75

Slide 74

When creating advanced reports, we often insert columns into our tables. We insert columns to add additional objects and formulas.

Insert a Column

1) Use the report from the previous page.
2) Duplicate the active report by right-clicking on the report tab and selecting duplicate report from the pop-up menu.
3) Click on the Portfolio Name column to select it.
4) Select *Insert Column After* from the Reporting toolbar.
5) Drag the Trans Year object from the Data tab and drop it in the center of a cell in the new column.

Inserting Columns (Data Tab)

- We can insert a column by dragging an object from the Data tab of the Manager. We drop the object on the right-side of a column in the table to insert a column to the right, or on the left-side of a column to insert a column to the left.
  - If the object is dropped on the center of a column in the table, then the object will replace the current object of the column.

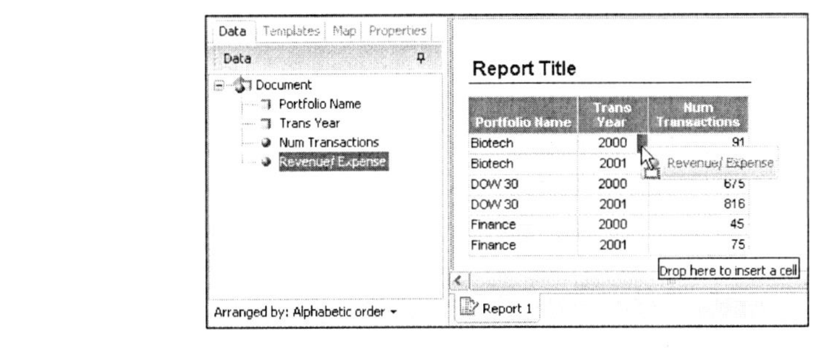

Slide 75

Objects can be dragged from the Data tab of the Manager to insert a column into a table. Be sure to drop the object on the right-side or the left-side of a cell in the column. If the object is dropped in the center of a cell in the column, then the dropped object will replace the current object in the column.

Insert a Column with the Data Tab

1) Use the report from the previous page.
2) Activate the duplicated report by clicking on the Report 1 (1) tab.
3) Click on the Trans Year object in the Data tab of the Manager and drag it to the right-side of any cell in the Portfolio Name column and release the mouse button.

Moving/Copying Tables

- We often move tables on a report. We move them around to make our reports more readable. We also copy tables so that we can convert them into a chart or other report structures. BusinessObjects allows us to quickly move tables, by clicking on the table's halo and then dragging it to a new position.
- We copy a table using the same method. However to copy the table to a new location, we press the [CTRL] key before releasing the mouse button.

Portfolio Name	Trans Year	Revenue/ Expens
Alternative Energy	2000	-89,221.3
Alternative Energy	2001	303,885.9
Biotech	2000	-369,888.4
Biotech	2001	449,314.1
DOW 30	2000	-4,385,412.5
DOW 30	2001	5,736,865.5

Portfolio Name	Trans Year	Revenue/ Expens
Alternative Energy	2000	-89,221.3
Alternative Energy	2001	303,885.9
Biotech	2000	-369,888.4
Biotech	2001	449,314.1
DOW 30	2000	-4,385,412.5
DOW 30	2001	5,736,865.5

Slide 76

Since tables can be transformed into any other report structure, we often copy tables to other locations on a report. Then, we transform them into other report components.

If we copy a table to just beneath another table, then there is a chance that the higher table may grow over the lower table when the document is refreshed. To prevent this scenario, we can place the lower table relative to the higher table. We will learn how to do this later in this chapter.

Move a Table

1) Move your cursor towards the outside edge of the default table.
2) When the grey halo appears, click and hold the left mouse button.
3) Move the table to a new location on the report.

Copy a Table

1) Follow the above steps, but before releasing the mouse button, hold down the [CTRL] key.

Properties Tab (Table)

- When a table is selected, the Properties tab in the Manager displays properties that we can set for the selected table.
- Many of the default properties for a table are sufficient. However, as we create more reports, we begin to develop our own styles and will quite often change the properties.
- The first property on the tab is the name. Sometimes it is convenient to name a structure, if there is more than one.
- The bottom of the Properties tab displays help on the currently selected option.

Data | Templates | Map | Properties

Properties

General	
Name	My Table
Display	
Cell spacing	0 px
Show table headers	Yes
Show table footers	Yes
Avoid duplicate row aggregation	Yes
Show rows with empty measure values	Yes
Show rows with empty dimension values	Yes
Show when empty	Yes
Appearance	
Background color	
Background image	
Borders	
Header cells	
Text Format	[Arial,9,Bold]
Background color	81, 117, 185
Background image	
Borders	
Body cells	
Text Format	[Arial,9,Regular]
Background color	255, 255, 255
Background image	
Borders	
Footer cells	
Text Format	[Arial,9,Regular]
Background color	255, 255, 255
Background image	
Borders	
Alternate Row/Column colors	
Frequency	2
Color	240, 240, 244

Display

Slide 77

The Properties tab of the Manager allows us to customize the appearance and behavior of the tables that we create. We rarely touch many of the options, but as we become more experienced report developers, we will find ourselves setting the properties more often.

We use the Properties of tables to enhance our reports. It is important to remember this is the purpose of the properties. Many beginner report developers will use properties to make their reports more pretty. While you can do this by setting certain properties, beautifying a report often makes it more difficult to read. Remember, we create reports so that people can better understand their business. If we make the report too beautiful, it will distract the viewer from this purpose.

When an option is selected, help text is displayed in the Descriptive area located below the Properties pane. This may help you to understand a property better.

Name a Table

1) Name the tables from the previous exercise. I named mine, *Top Table* and *Bottom Table.*

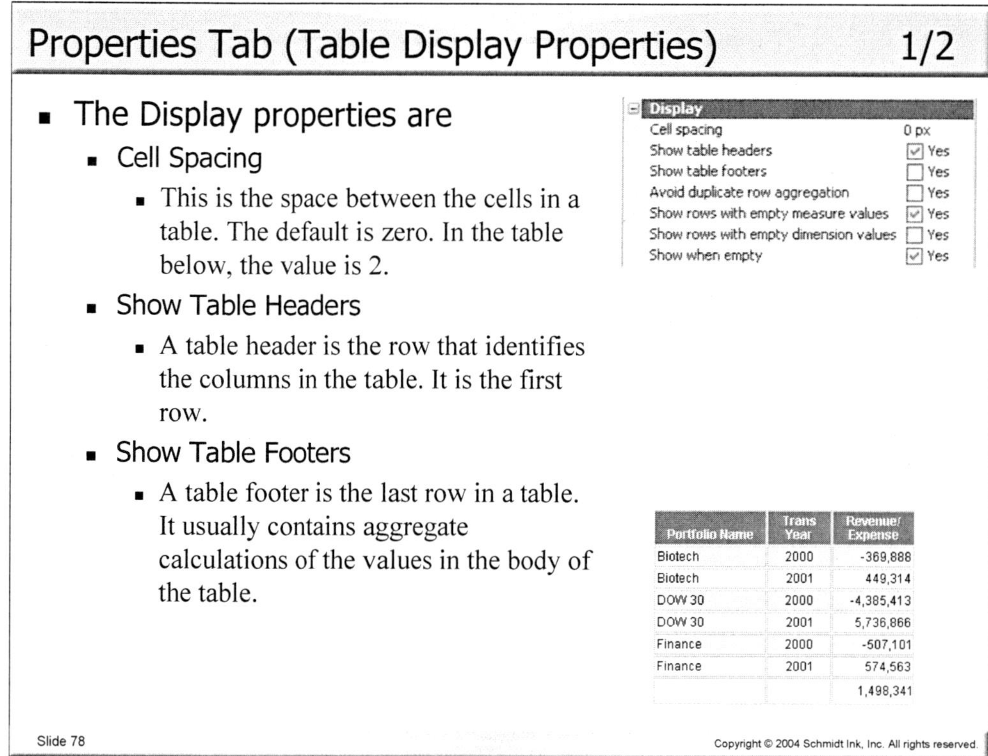

Properties Tab (Table Display Properties) 1/2
The Display properties are
Cell Spacing
This is the space between the cells in a table. The default is zero. In the table below, the value is 2.
Show Table Headers
A table header is the row that identifies the columns in the table. It is the first row.
Show Table Footers
A table footer is the last row in a table. It usually contains aggregate calculations of the values in the body of the table.
Display
Cell spacing 0 px
Show table headers Yes
Show table footers Yes
Avoid duplicate row aggregation Yes
Show rows with empty measure values Yes
Show rows with empty dimension values Yes
Show when empty Yes
Portfolio Name | Trans Year | Revenue/ Expense
Biotech | 2000 | -369,888
Biotech | 2001 | 449,314
DOW 30 | 2000 | -4,385,413
DOW 30 | 2001 | 5,736,866
Finance | 2000 | -507,101
Finance | 2001 | 574,563
1,498,341
Slide 78
Copyright © 2004 Schmidt Ink, Inc. All rights reserved.

Almost all tables have a header to identify the values in the columns beneath it. Headers are usually one row, but can be more. We can place measures and some formulas in the header. It is important to realize that since a header is singular in nature, and that only objects that display a single value can be placed in the header. For example, = Max([Date]), can be placed in the header. However, =[Date], usually can not, because it is a dimension that may represent more than one value.

Footers in tables are used to display totals. We usually aggregate the information in the rows above the footer. Most tables have some sort of calculation in the footer – Sum, count, average, max, min, and so forth.

Set the Cell Spacing and Show Headers and Footers

1) Select the table on the report.
2) Enter a value of 2 in the Cell Spacing option.
3) The Show Table Headers option is selected by default. You can un-check it and then check it back.
4) Check the Show Table Footers option.

Optional

5) Click on any value in the Revenue/Expense column and drag it to a cell in the footer. Before releasing the mouse button, hold down the [CTRL] key to copy the value to the footer and not move it.

Properties Tab (Table Display Properties) 2/2

- Avoid Duplicate Row Aggregation
  - Earlier, we learned that if we create a query with no measures, then Web Intelligence may return many duplicate rows to the document. However, Web Intelligence will roll-up the duplicates into unique rows. To stop this rolling up, check the *Avoid Duplicate Row Aggregation* option.
- Show rows with ...
  - These options allow you to hide rows or even the entire table if there are empty values. For example, if one of the rows in the table to the right, has no Revenue/ Expense value, then we could hide that row by un-checking the *Show rows with empty measure values* option.

Display	
Cell spacing	0 px
Show table headers	☑ Yes
Show table footers	☐ Yes
Avoid duplicate row aggregation	☐ Yes
Show rows with empty measure values	☑ Yes
Show rows with empty dimension values	☐ Yes
Show when empty	☑ Yes

Portfolio Name	Revenue/ Expense
Biotech	79,426
DOW 30	1,351,453
Finance	67,462

⇩

Portfolio Name	Revenue/ Expense
Biotech	449,314
Biotech	-369,888
DOW 30	-4,385,413
DOW 30	5,736,866
Finance	-507,101
Finance	574,563

Avoid Duplicate Row Aggregation

Slide 79

Sometimes the rows in a table are more aggregated than the rows in the query. This can happen for several reasons – a query can contain no measure objects, a dimension column can be deleted from a report (which would cause the rows to roll-up to the unique combinations of the remaining dimension values.) We rarely want to show these duplicate combinations on a report. However, sometimes we want to view the un-aggregated rows while we are debugging a report or trying to make sense of the totals. We use the Avoid Duplicate Row Aggregation option to un-aggregate these rows.

Sometimes there is no data in a query for certain levels of information. For example,

- A salesperson may not have made any sales for a certain time period. (Empty Measure)
- Sales may have been made, but no salesperson was assigned to the sale. (Empty Dimension)
- There may be no sales made by any salesperson. (Empty Table)

In some cases we want to hide the rows or tables that contain these situations. We can use the two Show Rows With... or the Show When Empty options to hide rows or tables that contain these empty values.

Show Duplicate Rows

1) Right-click on any value in the Trans Year column and select Remove Column from the pop-up menu.
   This will cause the rows to aggregate to the Portfolio Name.
2) Select the table, by clicking on the outmost border and then click on the Properties tab in the Manager to activate it.
3) Check the *Avoid Duplicate Row Aggregation* option to stop the table from rolling-up to Portfolio Name.

Properties Tab (Appearance Properties)

- The Appearance section of the Tables Properties allows us to set the colors, borders, and fonts of a table.
- The Appearance section contains five groups
  - Table, Header Cells, Body Cells, Footer Cells, and Alternate Row / Column colors.
  - From these sections, we can quickly format most of our table.
    - For example, most of the time I like to wrap the text and bottom align the table headers. This allows me to have descriptive headers, without taking too much report real estate.

Appearance	
Background color	255, 255, 255
Background image	Curve
Borders	
Header cells	
Text Format	[Arial,9,Bold]
Font name	Arial
Size	9
Style	Bold
Underline	Yes
Strikethrough	Yes
Text color	255, 255, 255
Wrap text	Yes
Vertical text alignment	Bottom
Horizontal text alignment	Center
Background color	81, 117, 185
Background image	
Borders	
Body cells	
Text Format	[Arial,9,Regular]
Background color	255, 255, 255
Background image	
Borders	
Footer cells	
Text Format	[Arial,9,Regular]
Background color	255, 255, 255
Background image	
Borders	
Alternate Row/Column colors	
Frequency	2
Color	240, 240, 244

Slide 80

The Appearance properties allow you to set the appearance of groups of cells. This is quicker then selecting them individually on a report and then setting the options. It also ensures that all cells will have the same appearance. For example, you can set the background color for body cells with the Appearance properties in the Table Properties tab. You can also select one column at a time and then set the background color. However, while you are setting them individually, you may select a slightly different color.

The background color for the table is only visible when there is an empty space. For example, when there is cell spacing. The table background color is behind the cells in the table

Set the Header Properties

1) Create a report with Portfolio Name, Trans Year, and Revenue/Expense
2) Select the entire table by clicking on the outermost border.
3) Click on the Properties tab of the Manager
4) Click the plus (+) sign in front of the Text Format option in the Header Cells section.
5) Check the Wrap Text option.
6) Select the bottom value for the Vertical Text Alignment option.
7) Select the center option for the Horizontal Text Alignment option.

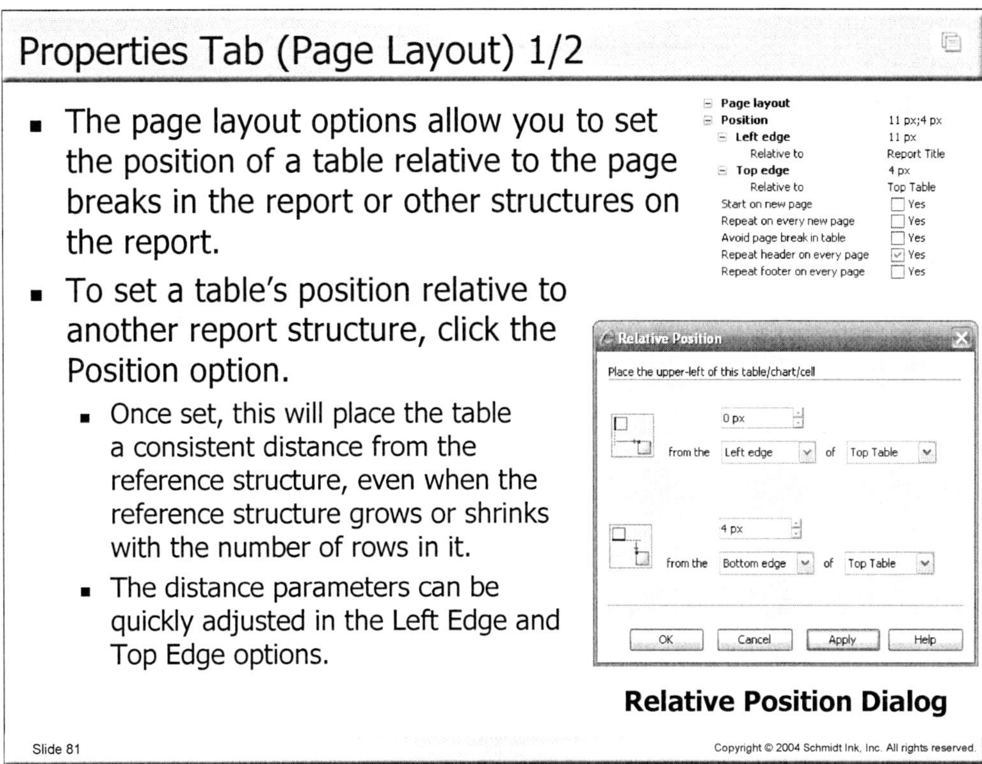

Properties Tab (Page Layout) 1/2
The page layout options allow you to set the position of a table relative to the page breaks in the report or other structures on the report.
To set a table's position relative to another report structure, click the Position option.
Once set, this will place the table a consistent distance from the reference structure, even when the reference structure grows or shrinks with the number of rows in it.
The distance parameters can be quickly adjusted in the Left Edge and Top Edge options.
Page layout
Position
11 px;4 px
Left edge
11 px
Relative to
Report Title
Top edge
4 px
Relative to
Top Table
Start on new page
Yes
Repeat on every new page
Yes
Avoid page break in table
Yes
Repeat header on every page
Yes
Repeat footer on every page
Yes
Relative Position
Place the upper-left of this table/chart/cell
0 px
from the
Left edge
of
Top Table
4 px
from the
Bottom edge
of
Top Table
OK
Cancel
Apply
Help
Relative Position Dialog
Slide 81
Copyright © 2004 Schmidt Ink, Inc. All rights reserved.

The Page Layout section allows a table to be placed on a report. Since most reports are dynamic in nature – each time it is refreshed, the number of data rows probably changes. This means that page breaks and other structures on the report will effect how a report is displayed. The Position option allows us to place a table on a report relative to other structures on the report, including the report page itself.

Place a Table Relative to Another

1) Create a Report with Portfolio Name, Trans Year, and Revenue Expense.
   A default table will be created after the query is ran.
2) Select the newly created table and name it Top Table in the Name property of the Properties tab.
3) Copy the table to just below itself
    1) Move the cursor to the topmost border of the table.
    2) When the cursor changes to a four way arrow, click and hold the mouse button down.
    3) Drag the table to just below itself, hold down the [CTRL] key, and release the mouse button.
4) Click on the edge of the table to select it.
5) Click on the Properties tab in the Manager.
6) Click the plus (+) sign preceding the Page Layout section to expand the section.
7) Move you mouse over the value portion of the Position option, when the values button appears, click on it.
8) Set the top field to 0 px, From the Left Edge, of Top Table.
9) Set the second field to 24 px, from the bottom, of Top Table.
10) Click Ok
    If you are unhappy with the distances, you can quickly adjust them by adjusting the values in the Left Edge and Top Edge options.

Properties Tab (Page Layout) 2/2

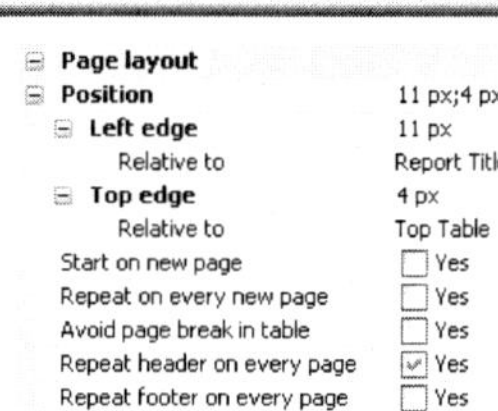

- When there is more than one table or a single table is longer than a single page, the page break options become very important to the appearance of a printed report.
- To view how a report will print, click the View Page Layout button in the Reporting toolbar.
- The available options are
  - Start on New Page. If the report has two tables, this option will start the second on a new page.
  - Repeat on every page. Repeats the table on every page.
  - Avoid page break in table. If there is more than one table and the second won't fit completely on a page, place it on a new page.
  - Repeat Header (Footer) on every page. A good option for the header, because headers identify the columns in a report.

Slide 82

If there is more than one table on a report or if the table spans many pages, the Page Layout options become very important, if we are to print the report. These options tell Web Intelligence what to do when a page break occurs in a table in a report. In most cases, we want the header of a table to repeat, because headers are needed to identify the columns in a report. Headers also assure viewers of the report that it is printing properly.

Table Templates

- Most reports start out as a table, since a table is the default report structure for most installations.
- Table templates allow us to transform tables (or any of the report structures) into any other types of report structures.
- To use a table template, click anywhere on the table and then activate the Templates tab by clicking on it. Then, simply drag a template icon unto the table.
- This table transformation is a very powerful technique, because it allows us to get the objects and formulas correct while in table format, then we simply convert the table to another type of report structure, such as a crosstab or a chart.

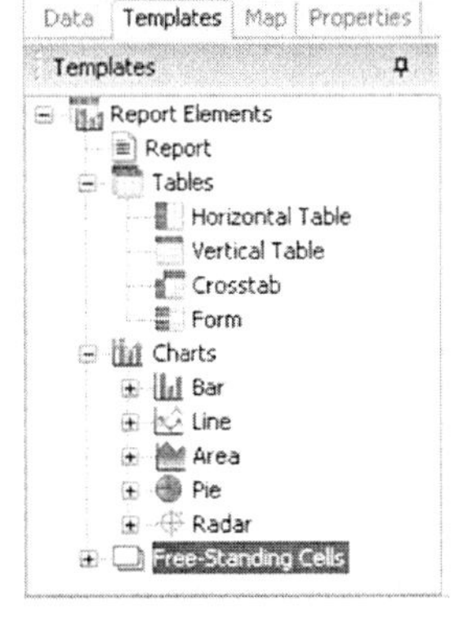

Slide 83

The table templates are very convenient, because they allow us to build a table with formulas and objects before we transform it into another type of report structure. This also allows us to copy a table and then transform the copy to a crosstab or a chart with the same information. Over the next few slides we will discuss some of the available templates.

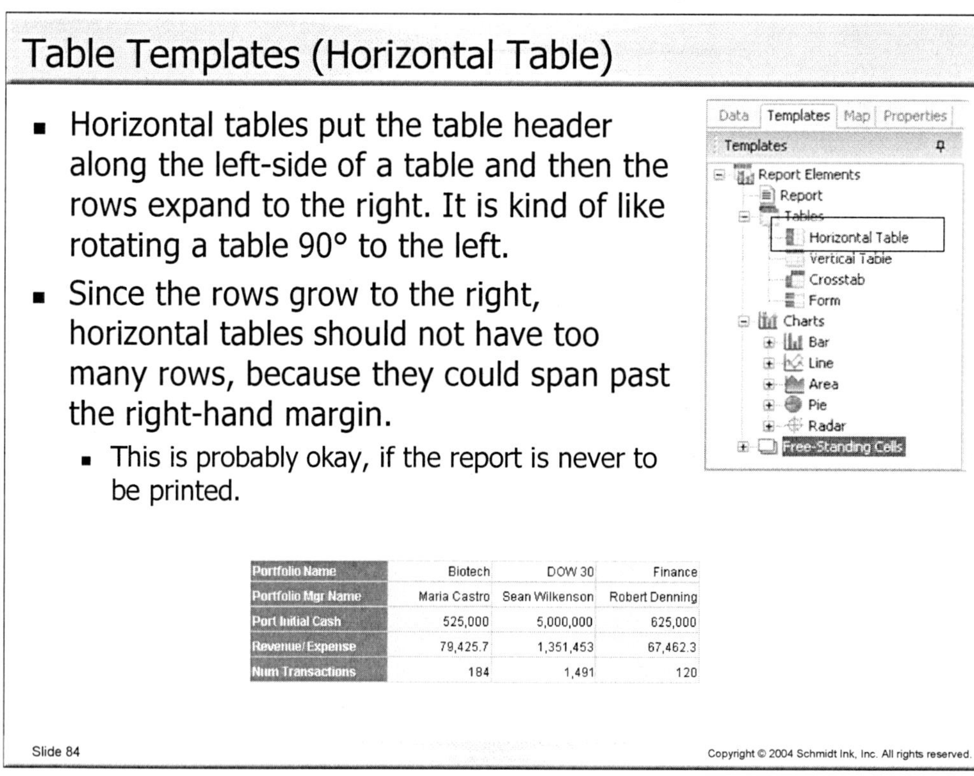

Table Templates (Horizontal Table)
Horizontal tables put the table header along the left-side of a table and then the rows expand to the right. It is kind of like rotating a table 90° to the left.
Since the rows grow to the right, horizontal tables should not have too many rows, because they could span past the right-hand margin.
This is probably okay, if the report is never to be printed.
Data
Templates
Map
Properties
Templates
Report Elements
Report
Horizontal Table
Vertical Table
Crosstab
Form
Charts
Bar
Line
Area
Pie
Radar
Free-Standing Cells
Portfolio Name
Biotech
DOW 30
Finance
Portfolio Mgr Name
Maria Castro
Sean Wilkenson
Robert Denning
Port Initial Cash
525,000
5,000,000
625,000
Revenue/Expense
79,425.7
1,351,453
67,462.3
Num Transactions
184
1,491
120
Slide 84
Copyright © 2004 Schmidt Ink, Inc. All rights reserved.

Horizontal tables place data side-by-side, not vertically as in traditional tables. This side-by-side placement allows the data to be easily compared. Thus these types of reports are usually used for data comparison, such as Profit and Loss statements. In fact, many people refer to this format as the financial format. However, it also works to compare down-time in factories, number of injuries per quarter, quarterly production increases, and so forth.

Create a Horizontal Table

1) Create a document with Portfolio Name, Portfolio Mgr Name, Port Initial Cash, Revenue/Expense, and Num Transactions.
2) Click anywhere on the default table to select it.
3) Click on the Templates tab to activate it.
4) Open the Tables class by clicking on the plus (+) sign preceding it.
5) Click on the Horizontal Table icon and drag it to the default table.

Table Templates (Vertical Table)

- Vertical tables are the default report structure for most installations of Web Intelligence.
  - We rarely use this template, unless we want to revert to a vertical table from a crosstab or other structure.
  - Vertical tables are the basic structure that we can convert into all other report structures.

Portfolio Name	Trans Year	Revenue/ Expense	Num Transactions
Alternative Energy	2000	-89,221	78
Alternative Energy	2001	303,886	96
Biotech	2000	-369,888	91
Biotech	2001	449,314	93
DOW 30	2000	-4,385,413	675
DOW 30	2001	5,736,866	816
Finance	2000	-507,101	45
Finance	2001	574,563	75
Media	2000	-465,394	62
Media	2001	386,854	58
Technology	2000	-1,116,767	274
Technology	2001	1,059,497	317

Slide 85

Vertical tables are the default report structure. We can add breaks to vertical tables or use them in section reports. They are the basic reporting structure.

Create a Horizontal Table

1) Create a document with Portfolio Name, Trans Year, Revenue/Expense, and Num Transactions.

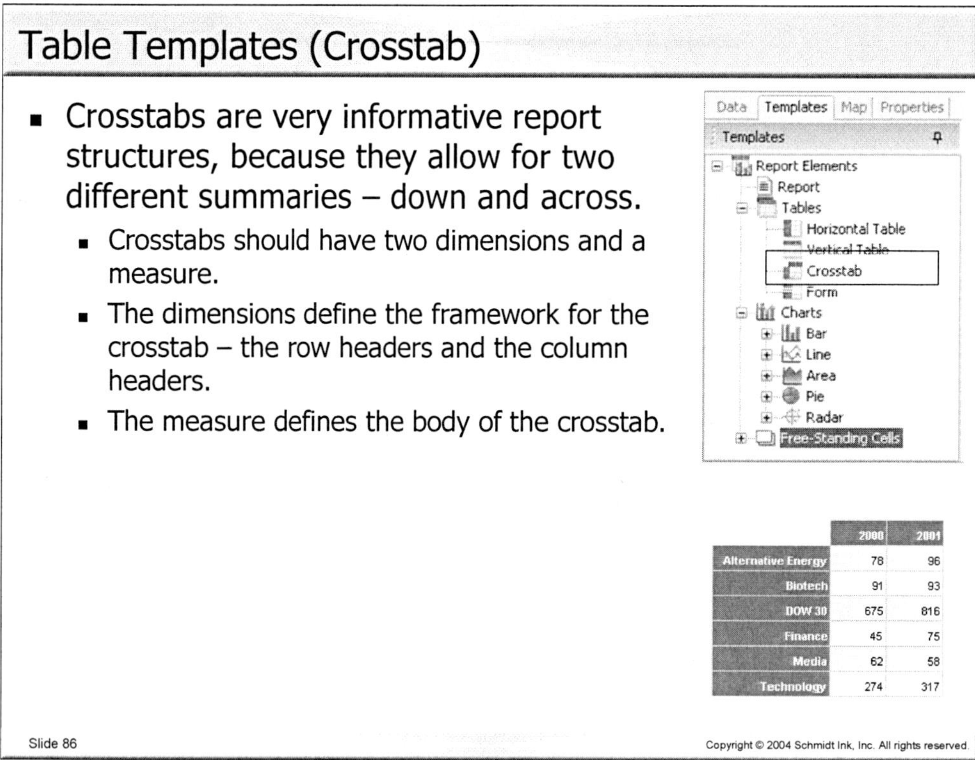

Crosstabs are one of the most common report structures, because they provide so much summary information in an optimal format. Many report readers appreciate the amount of information that may be presented in a crosstab.

Create a Crosstab

1) Create a document with Portfolio Name, Trans Year, and Num Transactions.
2) Drag the Crosstab template unto the default table report structure in the report.

Alternate Crosstab Method

- In the previous slide, we used the Crosstab table template to create a crosstab.
  - This method is sufficient in most cases, but the table must be organized so that the crosstab will have the proper row and column headers.
- We can create a crosstab by dragging a dimension from the table or the Data tab of the Manager to just above a table in a report.

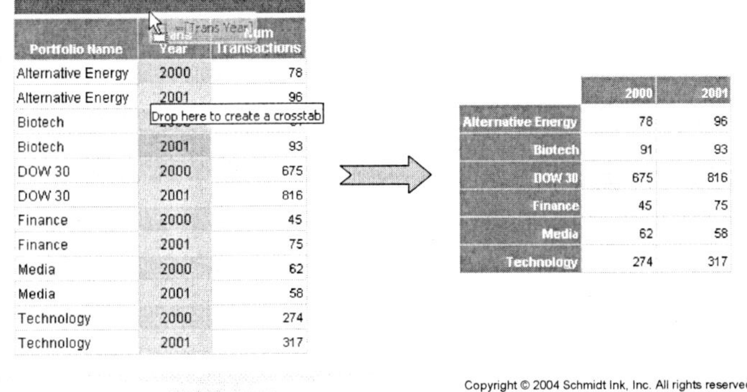

Portfolio Name	Trans Year	Num Transactions
Alternative Energy	2000	78
Alternative Energy	2001	96
Biotech	[illegible]	[illegible]
Biotech	2001	93
DOW 30	2000	675
DOW 30	2001	816
Finance	2000	45
Finance	2001	75
Media	2000	62
Media	2001	58
Technology	2000	274
Technology	2001	317

	2000	2001
Alternative Energy	78	96
Biotech	91	93
DOW 30	675	816
Finance	45	75
Media	62	58
Technology	274	317

Slide 87

Dragging and dropping dimension objects to create a crosstab is sometimes more efficient than using a template, because you get to decide which object will create the column header. You can also drag a measure object to create a crosstab, but this is usually not the case.

Create a Crosstab with Drag and Drop

1) Create a document with Portfolio Name, Trans Year, and Num Transactions.
2) Drag the Trans Year object to just above the table.
   Watch for the tooltip that tells you to drop to create a crosstab, because if you drop it in the wrong place, it will create a different type of report.

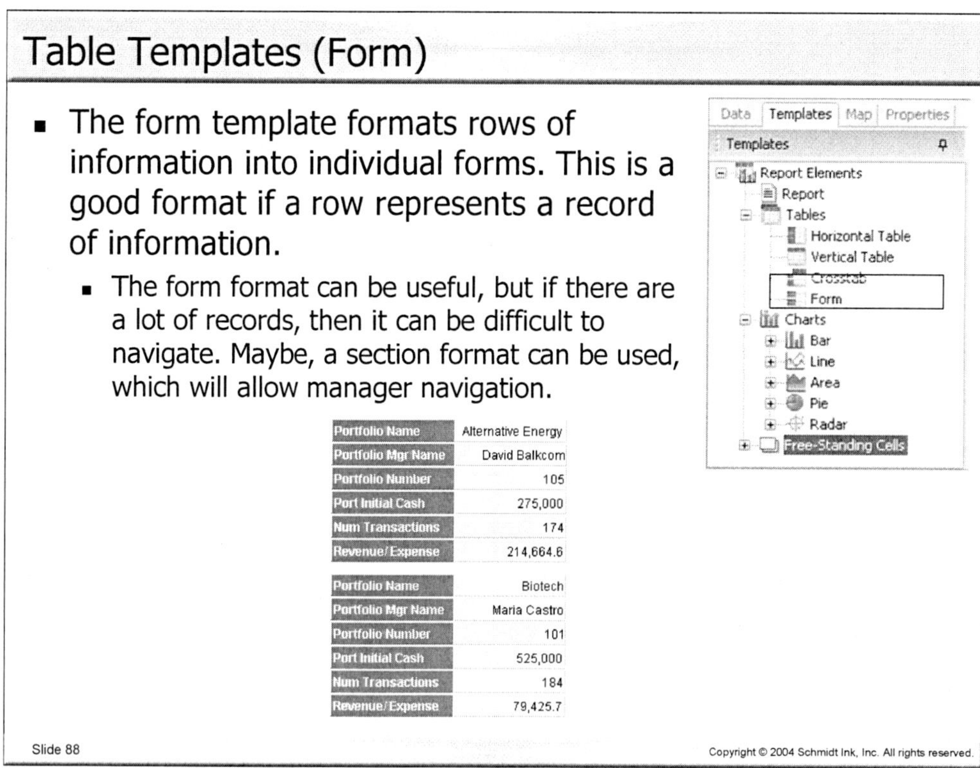

Forms are useful because they organize a row of data from the data provider into a single easy to read record. These types of reports are most useful if they are to be printed or contain relatively few records. If the report is to be viewed online, then it can be difficult to find a record quickly. If we use a section report and format the individual blocks to appear to use the form format, then we can use the Map tab of the Manager to quickly locate information. We will learn how to create a section report on the next slide.

Create a Form Report

1) Create a document with Portfolio Name, Portfolio Mgr Name, Portfolio Number, Port Initial Cash, Num Transactions, and Revenue/Expense.
2) Drag the Form template unto the default table report structure in the report.

Section Report

- Section reports divide a report into sections. Each section is a container for other report structures, such as tables, charts, and freestanding cells.
  - Each section is filtered by a value of the dimension used to create the sections.
- To create a section, drag a dimension from a table or the Manager and drop it on the report white-space.

=[Portfolio Name]

Drop here to create a section

Portfolio Name	Trans Year	Trans Quarter	Revenue/ Expense	Num Transactions
Biotech	2000	3	(400,996)	43
Biotech	2000	4	31,108	48
Biotech	2001	1	(119,552)	37
Biotech	2001	2	273,348	45
Biotech	2001	3	295,519	11
DOW 30	2000	3	(2,578,025)	322
DOW 30	2000	4	(1,807,388)	353
DOW 30	2001	1	330,350	369
DOW 30	2001	2	1,340,450	372
DOW 30	2001	3	4,066,066	75

Biotech

Trans Year	Trans Quarter	Revenue/ Expense	Num Transactions
2000	3	(400,996)	43
2000	4	31,108	48
2001	1	(119,552)	37
2001	2	273,348	45
2001	3	295,519	11

DOW 30

Trans Year	Trans Quarter	Revenue/ Expense	Num Transactions
2000	3	(2,578,025)	322
2000	4	(1,807,388)	353
2001	1	330,350	369
2001	2	1,340,450	372
2001	3	4,066,066	75

Slide 89

Section reports divide reports into sections. Each section is filtered by a value of the dimension that was used to create the section. In this example, the Portfolio Name dimension is used to create the sections. To create the sections, the dimension was dragged from the table and dropped on the report white-space. (White-space on a report is where no report structure exists) The report is then divided into sections created by each value of the dimension.

You can also right-click on a dimension in a table and select *Set as Section* from the pop-up menu to create a section report.

Create a Section Report

1) Create a document with Portfolio Name, Trans Year, Trans Quarter, Revenue/Expense and Num Transactions.
2) Drag the Portfolio Name object from the table and drop it onto the report white-space. (It is best to drop it above the table, because if you drop it on the side, then it will create sections, but the table may not be part of the sections)

Removing Sections

- To remove a section from a report, right-click on the white-space in a section and then select Remove from the pop-up menu.
  - The master cell can also be dragged back into the table or deleted from the report. A dialog box will be displayed with the option to remove the section from the report.
- Removing a section will not place the section dimension back in the table. To add the section dimension back in the table, drag it from the Data tab in the Manager.

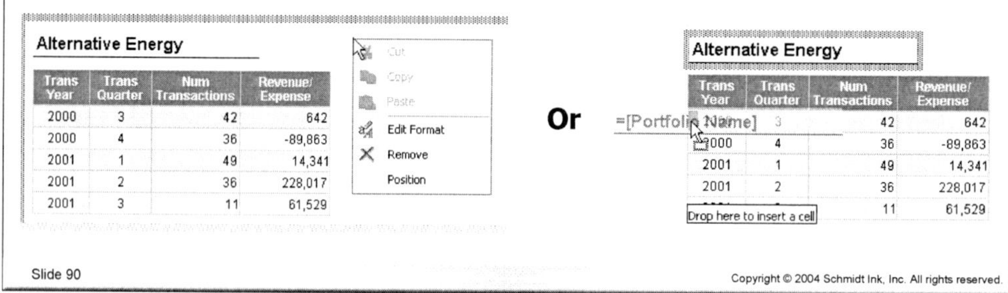

You can remove a section from a report by right-clicking on the section and then selecting Remove from the pop-up menu. You can also just remove the master cell from the section.

Remove a Section

1) Use the report created in the previous example
2) Right-click on the report tab and select Duplicate Report from the pop-up menu. (We are duplicating the report, because we will need the report for the next example)
3) Right-click on the master cell in a section and choose Remove from the pop-up menu.
4) When the dialog asking if you want to also delete the section is displayed, click on the Yes option.
5) Right-click on the report tab and select Remove Report from the pop-up menu. (We won't need it anymore)

Section Report (Multiple Report Structures)

- One of the great advantages of section reports is that each section can contain multiple report structures. The structures in the sections are all filtered by the section's dimension value (Sometimes referred to as the master cell).
- If a document contains multiple queries, and the section's dimension is merged, then the structures from the additional queries will also be filtered.

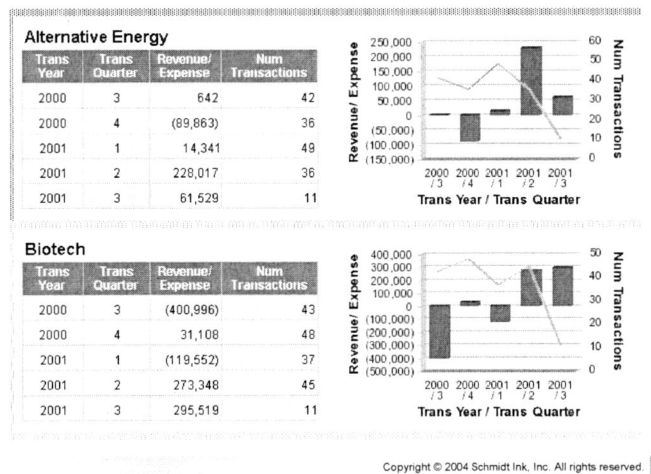

Alternative Energy

Trans Year	Trans Quarter	Revenue/ Expense	Num Transactions
2000	3	642	42
2000	4	(89,863)	36
2001	1	14,341	49
2001	2	228,017	36
2001	3	61,529	11

Biotech

Trans Year	Trans Quarter	Revenue/ Expense	Num Transactions
2000	3	(400,996)	43
2000	4	31,108	48
2001	1	(119,552)	37
2001	2	273,348	45
2001	3	295,519	11

Slide 91

Sections on a report can contain multiple report structures. In the above example, there is a table and a chart. Each structure in the section will be filtered by the section's dimension value. The section delimiters are visible when the report is clicked on, as shown in the example.

It's interesting that the report in the example is very powerful and yet only takes a few minutes to create. This is one of the true powers of Business Objects.

If a document contains more than one query and the query's have a dimension that is merged with the section's dimension object, then the structures from other queries will also be filtered by the section's master value. Many people use this technique to synchronize data from multiple data providers that are only merged on a single dimension common to all queries.

Create a Section with Multiple Report Structures

1) Use the report created in the previous example
2) Copy the table in the section to the right of the table.
3) Click on the Templates tab of the Manager.
4) Open the Charts group.
5) Open the Bar group in the Charts Group.
6) Drag the Vertical Bar and Line template onto the copied table.

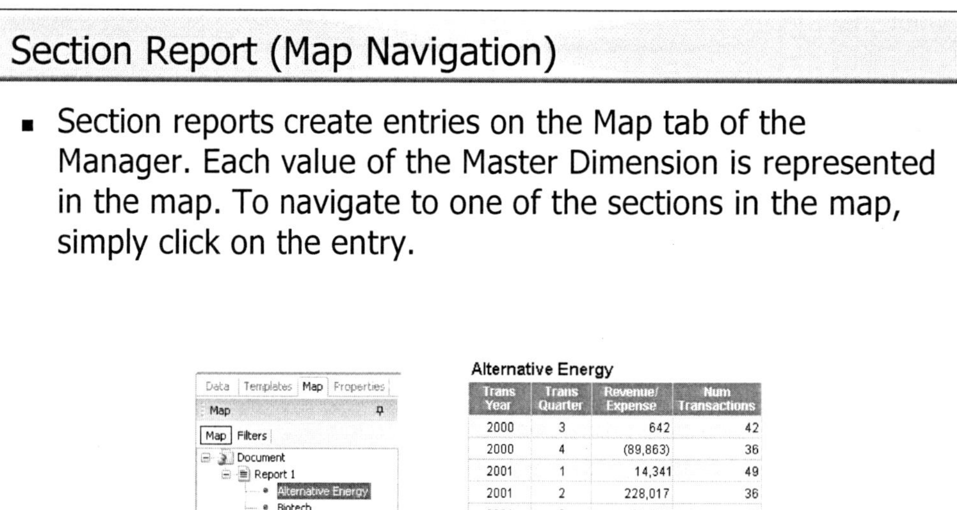
Section Report (Map Navigation)
▪ Section reports create entries on the Map tab of the Manager. Each value of the Master Dimension is represented in the map. To navigate to one of the sections in the map, simply click on the entry.
Data
Templates
Map
Properties
Map
Map
Filters
Document
Report 1
Alternative Energy
Biotech
DOW 30
Finance
Media
Technology

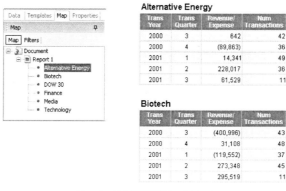

Alternative Energy

Trans Year	Trans Quarter	Revenue/ Expense	Num Transactions
2000	3	642	42
2000	4	(89,863)	36
2001	1	14,341	49
2001	2	228,017	36
2001	3	61,529	11

Biotech

Trans Year	Trans Quarter	Revenue/ Expense	Num Transactions
2000	3	(400,996)	43
2000	4	31,108	48
2001	1	(119,552)	37
2001	2	273,348	45
2001	3	295,519	11

Slide 92

Another great advantage to Section reports is that they are very easy to navigate. This is a great advantage over the Form template, because the Form template does not create entries in the Map tab of the Manager.

Navigate a Section Report

1) Use the report created in the previous example
2) Click on the Map tab to activate it.
3) Click on any of the entries to navigate to a section.

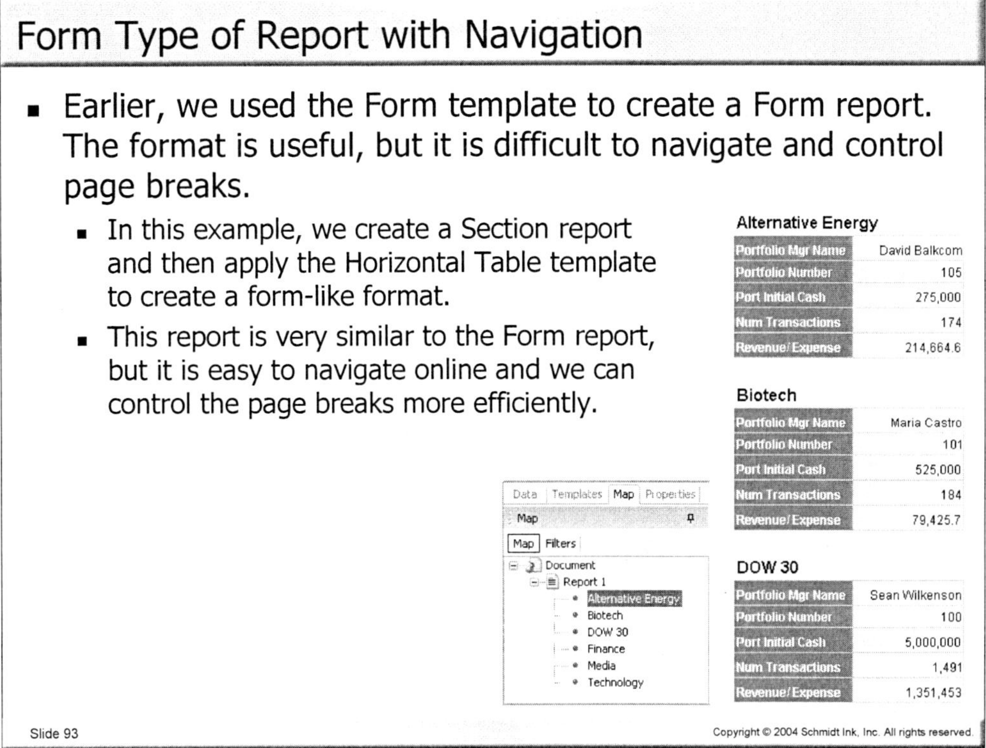

The Form template is useful, because it arranges the information into records. However, if the Form report is viewed online, then it is difficult to navigate because we must scroll to desired forms in the report. Therefore, it maybe better to create sections on the report that represent the values of a mapping dimension. Then we can use these values as an index on the Map tab of the Manager. If we rotate the table in each section, using the Horizontal Table template, then the tables will very much resemble the forms in the Form report.

Create a Section Form Report

1) Create a report with Portfolio Name, Portfolio Mgr Name, Portfolio Number, Port Initial Cast, Num Transactions, and Revenue/Expense.
   (These are the same objects used in the Form Template example)
2) Create Sections on the report by dragging the Portfolio Name object from the table and dropping it into the white-space above the table.
3) Click on the Templates tab of the Manager to activate it.
4) Drag the Horizontal Table template and drop it onto the table in a section.
5) View the Map tab and see that we can navigate the forms using the section map.

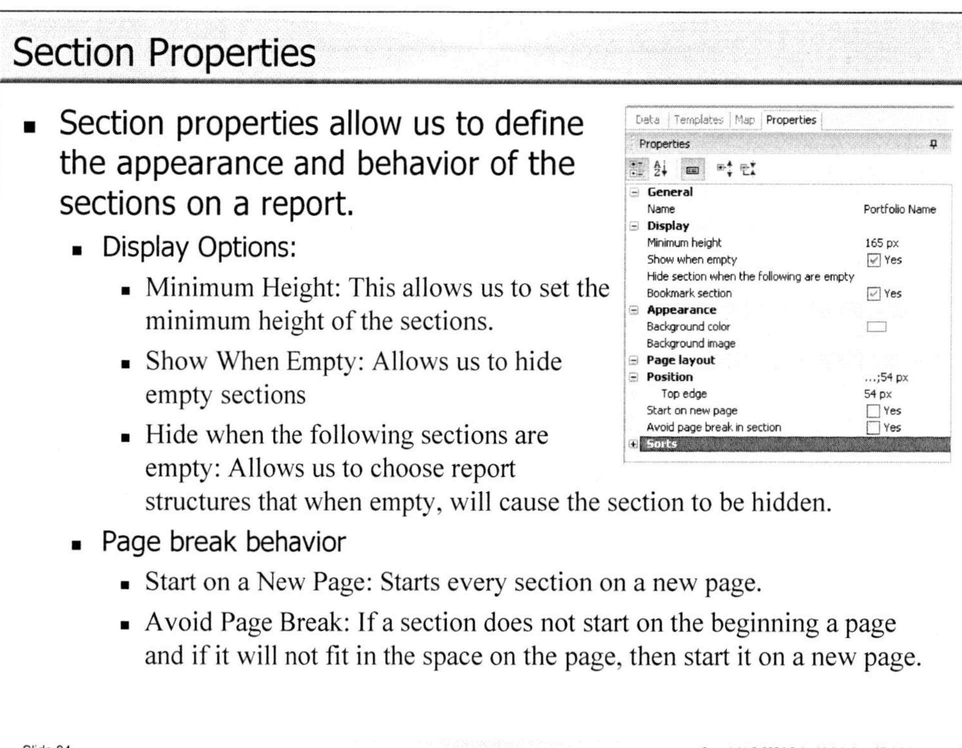

Section Properties
Section properties allow us to define the appearance and behavior of the sections on a report.
Display Options:
Minimum Height: This allows us to set the minimum height of the sections.
Show When Empty: Allows us to hide empty sections
Hide when the following sections are empty: Allows us to choose report structures that when empty, will cause the section to be hidden.
Page break behavior
Start on a New Page: Starts every section on a new page.
Avoid Page Break: If a section does not start on the beginning a page and if it will not fit in the space on the page, then start it on a new page.
Data
Templates
Map
Properties
Properties
General
Name
Portfolio Name
Display
Minimum height
165 px
Show when empty
Yes
Hide section when the following are empty
Bookmark section
Yes
Appearance
Background color
Background image
Page layout
Position
...;54 px
Top edge
54 px
Start on new page
Yes
Avoid page break in section
Yes
Sorts
Slide 94
Copyright © 2004 Schmidt Ink, Inc. All rights reserved.

The Section properties allow us to define how a section will appear and how it will behave when it encounters a page break. To see a section's properties, click anywhere on a section's white-space. Then click on the Properties tab of the Manager. The Page break options will have no effect, unless the report is placed into Page Layout mode or the report is printed.

These properties offer a great advantage over the Form report discussed earlier in this chapter.

View a Section's Properties

1) Create a report with Portfolio Name, Trans Year, Trans Quarter, Num Transactions, and Revenue/Expense.
2) Drag the Portfolio Name object from the table and drop it into the white-space above the table.
3) Click anywhere in the section's white space.
4) Click on the Properties tab of the Manager.

Inserting Breaks

- Breaks in a table or crosstab allow for subtotals without creating sections in a report.
- Differences between Sections and Breaks
  - Breaks do not create entries in the Map tab of the Manager.
  - Breaks allow for grand totals at the bottom of a table.
  - Sections create report wide sections; breaks are local to the table.
  - Break headers are more flexible.

Portfolio Name	Trans Year	Trans Quarter	Num Transactions	Revenue/ Expense
Biotech	2000	3	43	-400,996
Biotech	2000	4	48	31,108
Biotech	2001	1	37	-119,552
Biotech	2001	2	45	273,348
Biotech	2001	3	11	295,519
DOW 30	2000	3	322	-2,578,025
DOW 30	2000	4	353	-1,807,388
DOW 30	2001	1	369	330,350
DOW 30	2001	2	372	1,340,450
DOW 30	2001	3	75	4,066,066

Portfolio Name	Trans Year	Trans Quarter	Num Transactions	Revenue/ Expense
Biotech	2000	3	43	-400,996
	2000	4	48	31,108
	2001	1	37	-119,552
	2001	2	45	273,348
	2001	3	11	295,519
Biotech				

Portfolio Name	Trans Year	Trans Quarter	Num Transactions	Revenue/ Expense
DOW 30	2000	3	322	-2,578,025
	2000	4	353	-1,807,388
	2001	1	369	330,350
	2001	2	372	1,340,450
	2001	3	75	4,066,066
DOW 30				

Slide 95

Breaks in tables create groups within the table. These groups then can have their own headers and footers that can contain calculations that work on the values in the break. They are similar to Section reports, but they do not create report sections that span the width of the report. Breaks also do not create entries in the Map tab of the Manager.

Since breaks do not create entries in the Map tab, breaks are often used for printed reports. Since we can format them to have only one header per page and they do not have a master cell, break reports use vertical real-estate more efficiently. We will do this example on the following slide.

Create a Break Report

1) Create a document with Portfolio Name, Trans Year, Trans Quarter, Revenue/Expense and Num Transactions.
2) Click on any value in the Portfolio Name column and click the Insert/Remove Break button to insert the break.

Formatting Breaks

- Breaks have several formatting options
  - Show Break Header (Footer): Since every break has its own header and/or footer we can display these independently of the table header/footer.
  - We can remove the duplicate values from the break column (default): This makes each break value show once, as shown in the example.
  - Page break options allow us to control how the break will behave when it encounters a page break. The *Avoid page breaks in table* and *Repeat header on every page* are usually selected for most break reports.
    - Page break options are only useful when a report is printed or viewed in Page Layout mode.

Breaks

Show break header	☐ Yes
Show break footer	☑ Yes
Remove duplicate values	☑ Yes
Center values across break	☐ Yes
Apply implicit sort to values	☑ Yes

Page layout

Start on new page	☐ Yes
Avoid page breaks in table	☐ Yes
Repeat header on every page	☐ Yes
Repeat break value on new page	☐ Yes

Portfolio Name	Trans Year	Trans Quarter	Num Transactions	Revenue/ Expense
Biotech	2000	3	43	-400,996
	2000	4	48	31,108
	2001	1	37	-119,552
	2001	2	45	273,348
	2001	3	11	295,519
Biotech				
DOW 30	2000	3	322	-2,578,025
	2000	4	353	-1,807,388
	2001	1	369	330,350
	2001	2	372	1,340,450
	2001	3	75	4,066,066
DOW 30				

Slide 96

Most Breaks reports must be formatted if the report is to be printed or viewed in Page Layout mode. The formats allow us to customize how the breaks will behave.

Format a Break Report

1) Using the Break report from the previous example...
2) Click on the halo on the edge of the table to select the entire table.
3) Click the Properties tab in the Manager.
4) Open the Display section and select Yes for the *Show Table Headers* option. (We now have two headers at the top of our table.)
5) Click on any break value in the Portfolio Name column.
6) Click on the Properties Tab to activate it.
7) Open the Breaks section and clear the Yes option for the *Show Break Header* option.
8) Open the Page Layout section in the Breaks section.
9) Click place a check in the Yes option for the Avoid Page Breaks in table section.

Basic Aggregate Calculations in Tables

- We can add basic aggregate calculations to a table by selecting a value column in a table and then selecting an aggregate function from the Reporting toolbar.
  - If a break is added to a table, the calculation can be applied to all the break footers by re-clicking on the aggregate function in the Reporting toolbar.

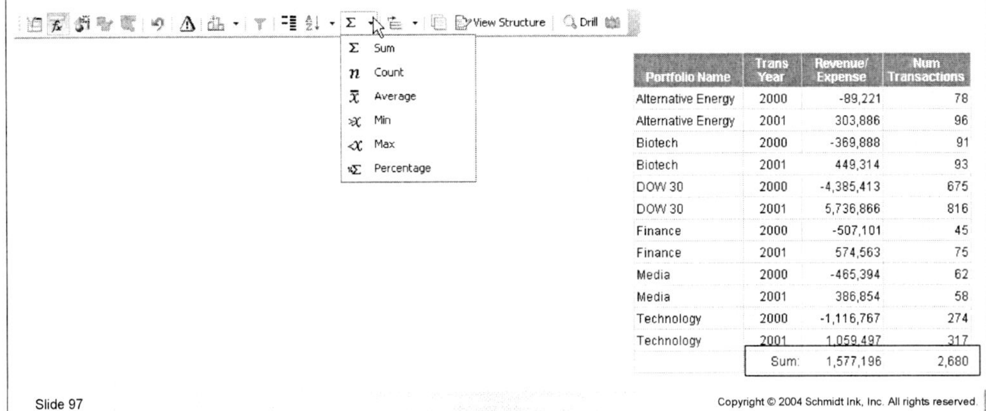

Portfolio Name	Trans Year	Revenue/ Expense	Num Transactions
Alternative Energy	2000	-89,221	78
Alternative Energy	2001	303,886	96
Biotech	2000	-369,888	91
Biotech	2001	449,314	93
DOW 30	2000	-4,385,413	675
DOW 30	2001	5,736,866	816
Finance	2000	-507,101	45
Finance	2001	574,563	75
Media	2000	-465,394	62
Media	2001	386,854	58
Technology	2000	-1,116,767	274
Technology	2001	1,059,497	317
	Sum:	1,577,196	2,680

Slide 97

Aggregate functions can be added to footers in a table by selecting any cell in a column and then selecting an aggregate function from the reporting toolbar.

Add Aggregate Functions to a Table

1) Create a report using Portfolio Name, Trans Year, Num Transactions, and Revenue/Expense.
2) Click on any value in the Revenue/Expense column and then click the Sum function in the Reporting toolbar.
3) Click on any value in the Num Transactions column and then click the Sum function in the Reporting toolbar.

Add a Break to the Table

4) Click on any Portfolio Name value and click the Insert/Remove Break button (Notice that there are no calculations in the break footers.)
5) Click on any Revenue/Expense value and click the Sum function on the Reporting toolbar
6) Click on any Num Transactions value and click the Sum function on the Reporting toolbar (This will add calculations to the break footers.)

Default Measure Aggregate Calculations

- Each measure (usually) has a default calculation associated with it. This is the definition of a measure.
- We can take advantage of these default calculations to simply our reports and make our report creation more efficient.
- The default calculation for Revenue/Expense and Num Transactions is Sum. Therefore, when the measure is placed in a footer or other context in a report, it will automatically sum (without the sum function explicitly in the formula)

Portfolio Name	Trans Year	Revenue/ Expense	Num Transactions
Alternative Energy	2000	-89,221	78
Alternative Energy	2001	303,886	96
Biotech	2000	-369,888	91
Biotech	2001	449,314	93
DOW 30	2000	-4,385,413	675
DOW 30	2001	5,736,866	816
Finance	2000	-507,101	45
Finance	2001	574,563	75
Media	2000	-465,394	62
Media	2001	386,854	58
Technology	2000	-1,116,767	274
Technology	2001	1,059,497	317
	Sum:	6	2,680

=[Revenue/ Expense]

Slide 98

Most measures have an aggregate function associated with them. This is the definition of a measure. Since this is true, we can just place the measure in various contexts of a table and the measure will conform to the context by aggregating all values in the context.

In the above example, there are two contexts – the individual rows (Defined by the Portfolio Name and Trans Year objects) and the table footer (Report level). The same formula can be used for all contexts, so to quickly add a sum, we can just drag and drop a measure into the footer of the table.

Use the Default Measure Calculation in the Table Footer

1) Create a report using Portfolio Name, Trans Year, Num Transactions, and Revenue/Expense.
2) Click on the halo around the table to select it.
3) Click on the properties tab to activate it.
4) Open the Display group on the Properties tab.
5) Check Yes on the *Show Table Footers* option.
6) Click on any value in the Num Transactions column in the table.
7) Drag the selection down to the newly visible table footer. Before releasing the mouse button to drop the object, hold down the [Ctrl] key to copy it to the footer.
   (If the [Ctrl] is not held down, then the object will be removed from the table and placed in the footer.)
   (Make sure that you drop the object in the center of the footer cell.)

Chart Templates

- With Web Intelligence, we can easily turn any table or crosstab into a chart.
  - Just select a chart template and drag it onto a table in a report.
- We can pick from five chart categories
  - Bar: Usually for discrete data, such as comparing sells of different products or regional revenue.
  - Line: Usually for continuous data, such as sells across time (months, quarters, years)
  - Area: Usually for volume data, such as sells across regions.
  - Pie: Usually one single dimensional data where distribution is significant, such as units of each product sold or regional revenue.

Data | Templates | Map | Properties

Templates

- Charts
  - Bar
    - Vertical Grouped
    - Horizontal Grouped
    - Vertical Stacked
    - Horizontal Stacked
    - Vertical Percent
    - Horizontal Percent
    - 3D Bar
    - Vertical Bar and Line
    - Horizontal Bar and Line
  - Line
    - Vertical Mixed
    - Horizontal Mixed
    - Vertical Stacked
    - Horizontal Stacked
    - Vertical Percent
    - Horizontal Percent
    - 3D Line
    - 3D Surface
  - Area
    - Vertical Absolute
    - Horizontal Absolute
    - Vertical Stacked
    - Horizontal Stacked
    - Vertical Percent
    - Horizontal Percent
    - 3D Area
    - 3D Surface
  - Pie
    - Pie
    - Doughnut
    - 3D Pie
    - 3D Doughnut
  - Radar

Slide 99

Selecting the right chart for data is almost as important as the chart itself. Charts must be easy to understand and clearly show the results of interest. For example, suppose you have a car dealership. You sell several different models and each salesperson can sell any of the models. You create a report that has a table with salesperson, model, and number of cars sold. This data would be perfect for a Bar chart, because it is discrete.

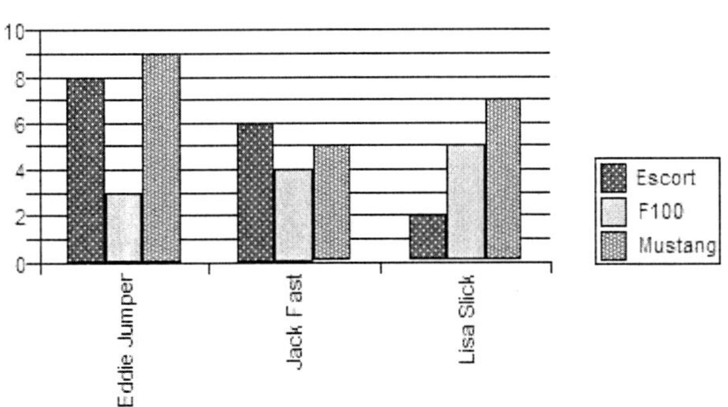

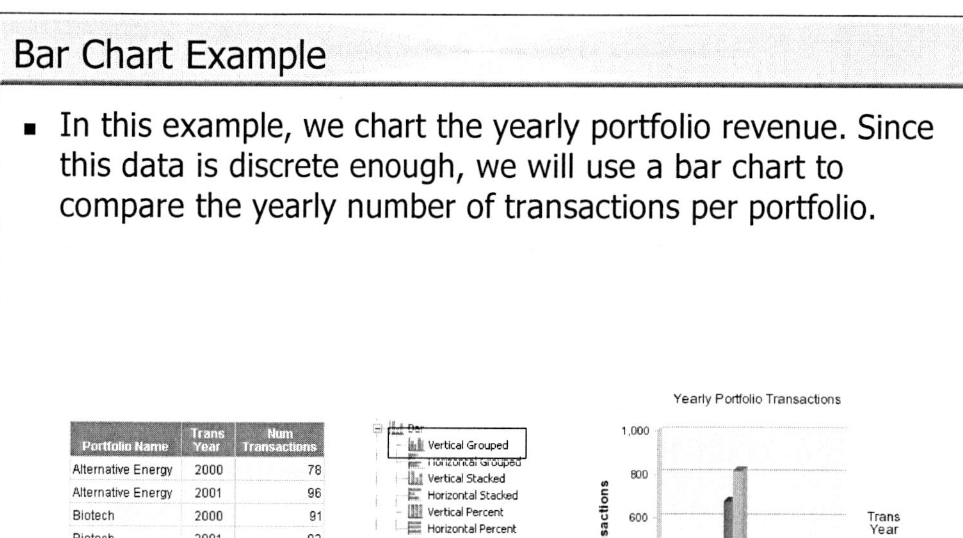

Portfolio Name	Trans Year	Num Transactions
Alternative Energy	2000	78
Alternative Energy	2001	96
Biotech	2000	91
Biotech	2001	93
DOW 30	2000	675
DOW 30	2001	816
Finance	2000	45
Finance	2001	75
Media	2000	62
Media	2001	58
Technology	2000	274
Technology	2001	317

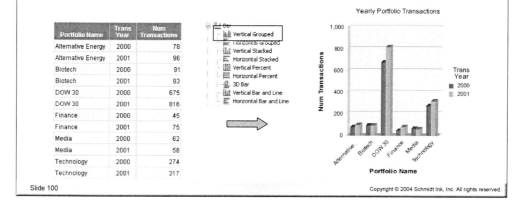

In this example we will use the Vertical Grouped Bar chart to compare the yearly number of transactions in each portfolio.

Create a Vertical Grouped Bar Chart

1) Create a report using Portfolio Name, Trans Year, and Num Transactions.
2) Click on the Templates to activate it.
3) Open the Charts group.
4) Open the Bar group.
5) Click on and drag the Vertical Grouped template unto the table in the report.

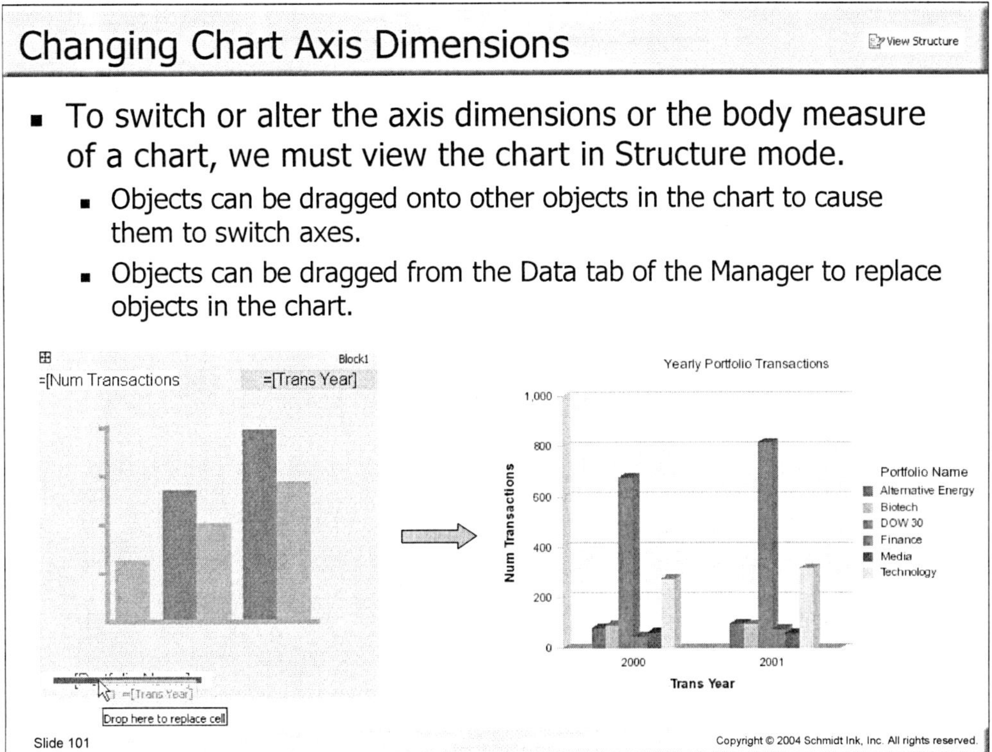

When we want to alter the dimensions in a table, we just click on them directly. However, in a chart the dimensions are not available in this manner. To access a chart's objects, we view the report in structure mode. Structure mode makes all objects and formulas in a chart accessible to be modified or replaced.

Changed the X-Axis Dimension with the Z-Axis Dimension

1) Create a report using Portfolio Name, Trans Year, and Num Transactions.
2) Click on the Templates to activate it.
3) Open the Charts group.
4) Open the Bar group.
5) Click on and drag the Vertical Grouped template unto the table in the report.
6) Click the View Structure button on the Reporting toolbar.
7) Click on the Trans Year object in the Chart Structure and drag it onto the Portfolio Name object.
(Make sure that you drop it in the center of the object.)
8) Click on the View Results (Formerly View Structure) button to see the results of the swap.

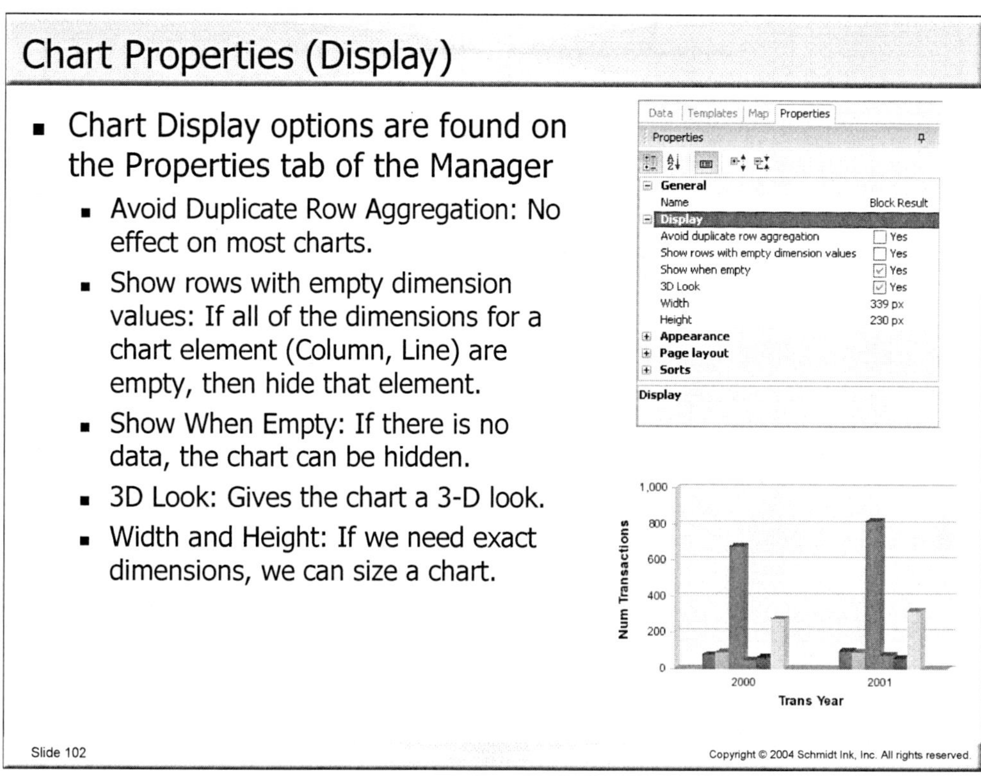
Chart Properties (Display)
Chart Display options are found on the Properties tab of the Manager
Avoid Duplicate Row Aggregation: No effect on most charts.
Show rows with empty dimension values: If all of the dimensions for a chart element (Column, Line) are empty, then hide that element.
Show When Empty: If there is no data, the chart can be hidden.
3D Look: Gives the chart a 3-D look.
Width and Height: If we need exact dimensions, we can size a chart.
Data
Templates
Map
Properties
Properties
General
Name
Block Result
Display
Avoid duplicate row aggregation
Yes
Show rows with empty dimension values
Yes
Show when empty
Yes
3D Look
Yes
Width
339 px
Height
230 px
Appearance
Page layout
Sorts
Display
1,000
800
600
400
200
0
Num Transactions
2000
2001
Trans Year
Slide 102
Copyright © 2004 Schmidt Ink, Inc. All rights reserved.

The display options for a chart allow us to define how a chart will be displayed.

Set Some Display Options

1) Create a Report with Portfolio Name, Trans Year, and Num Transactions.
2) Make the default table a crosstab, by dragging the Trans Year object from the table and dropping it to just above the table.
3) Drag the Vertical Grouped template from the Charts>Bar template group and drop it on the crosstab.
4) Click anywhere on the chart to make sure that it is selected.
5) Click the Properties tab in the Manager to activate it.
6) Clear the *Show When Empty* option.
7) Clear the *3D Look* option.
   (Look at the 2D Chart)
8) Re-check the 3D Look option.

Chart Properties (Appearance)

- Chart Display options are found on the Properties tab of the Manager
  - Color: Background and Wall: We can set the chart frame colors.
  - Borders: Set visible, color and width of chart border.
  - Show Floor: Yes to show.
  - Legend: Yes to show.
    - Position: We can select right, left or bottom.
    - Values: We can format the text, background color, and/or borders.
  - Title: Yes to show
    - Text: Enter text for the title
    - We can format the text, background color, and/or borders.

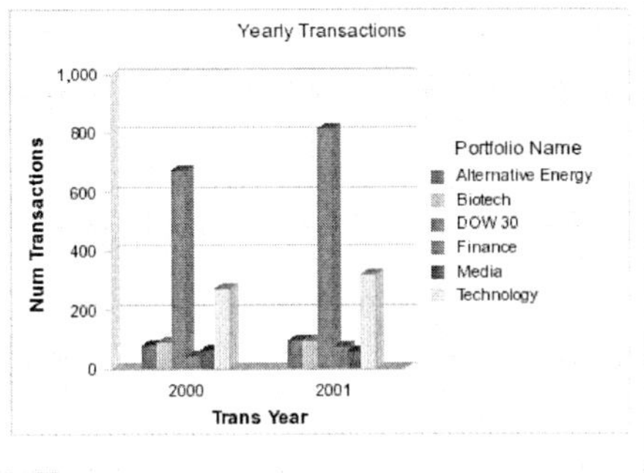

Slide 103

The Appearance options for a chart allow us to define how a chart will appear. If you are viewing this slide on a computer, the background color is light blue and the wall color is light yellow.

Set Some Appearance Options

1) Click on the Chart created in the previous example to select it.
2) Set the *Background Color* option to light blue.
3) Place your mouse over the cell to the right of the Borders option, then click on the small button that will appear. Set the border to medium width and light gray.
4) Set the *Wall Color* option to light yellow.
5) Check the *Yes* option for *Legend* to display the legend.
6) Choose the Right value for the *Position* option.
7) Select the *Yes* option for the Title.
8) Enter - Yearly Transactions – into the *Text* option.

Chart Properties (Appearance) - Data

- The Palette option allows us to pick the colors for the chart elements, such as bars, lines, or pie pieces.
- The *Show Data* option in the Appearance>Data section allows us to display values on the chart elements.
  - In some cases this is very useful, but in others it is not practical because the value labels may be too crowded.
  - The data labels in a 3D Look chart are very helpful, because they allow us to know the magnitude of each column without referring to the grid.
  - Sometimes it is good to format the values to have a white background to allow them to be easily read.

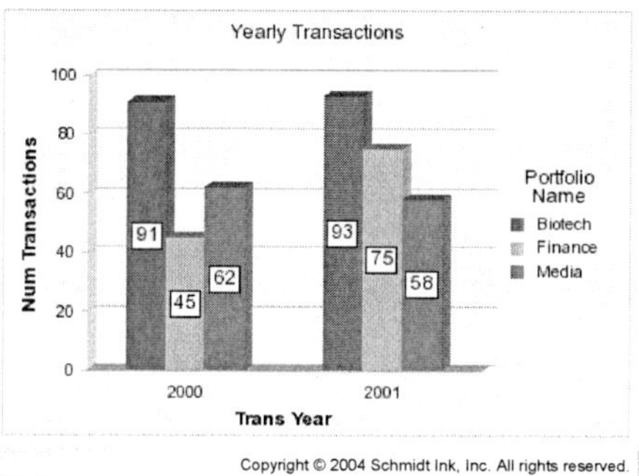

Slide 104

Sometimes we want to display the data values on the chart elements, as this makes it easier to determine the element's exact value. Most of the time, when we do display the values, we have to format them to make them readable. In this example, we set the background to white. This gives a constant background color so the values will always have the same contrast and therefore are easier to read.

Format the Data

1) Click on the Chart created in the previous example to select it.
2) Place your cursor over the cell to the right of the *Palette* option, and then click on the small button that appears.
   Choose a palette for your chart colors.
3) Check the *Yes* option for the *Show Data* in the *Values* group.
4) Open the *Text Format* option and set the Font *Size* to 10.
5) Set the Background color to white.
   (It may appear to be white, but it may be clear)
6) Place your cursor over the cell to the right of the *Borders* option and click on the small button that appears. Set the border to thin and black.
7) Set a filter to show only Biotech, Finance, and Media
   1) Click the Show/Hide Filter Pane on the Reporting toolbar
   2) Drag the Portfolio Name object into the Filter pane
   3) When the Filter Editor is displayed, select
   4) In List from the Operator drop list
   5) Value(s) from list from the Operand Type section
   6) Double-click Biotech, Media, and Finance to move them into the Value(s) Selected list.

Chart Properties (Appearance) - (X-Axis)

- The X-axis of a chart is usually the visible axis that has dimension values. In a Pie chart it is the Legend.
  - Grid: The Grid for the X-axis is the lines that run perpendicular to the axis at each dimension value.
  - Values: We can show/Hide the X-axis values
    - Number Format: We can format the numbers, dates, or Boolean values.
  - Label: The X-axis description.
    - Other label: We can change the label for the axis.
    - Text Format: We can format the font for the Label.
    - Borders: We can put a border around the label.

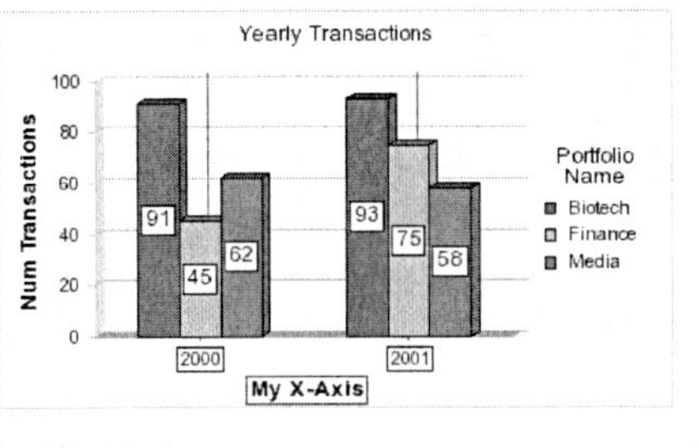

Slide 105

Sometimes we have to format the X-axis. We may want to override the label that Web Intelligence has assigned to it. We often do this when the object name is too abstract. We also have to format the values on the axis from time to time. For example, suppose the years were formatted as 2,000.00 (#,##0.00). We would have to format the number to 2000 (0). We can also format the font of the label or the values and add a border to them.

Format the X-Axis

1) Click on the Chart created in the previous example to select it.
2) Check the *Yes* option for the Grid.
3) Change the grid color to red.
   Place your cursor over the cell to the right of the *Grid Color* option.
   When the small button appears in the cell, click on it.
   Double-click on color when the palette is displayed.
4) Place borders around the axis values
   Place your cursor over the cell to the right of the Borders option in the Values section.
   When the small button appears, click on it.
   Select a style, color, and click the Box border button.
   Click Ok.
5) Change the axis label
   Enter My X-Axis into the Other Label option in the Label section.
6) Place a border around the label
   Place your cursor over the cell to the right of the Borders option in the Values section.
   When the small button appears, click on it.
   Select a style, color, and click the Box border button.
   Click Ok.

Chart Properties (Appearance) - (Y-Axis)

- The Y-axis of a chart measures the magnitude of measures. It is usually perpendicular to the X-axis
  - Grid: The Grid for the Y-axis is the lines that run perpendicular to the axis at each measure value (tick mark on the axis.)
  - Values: The same as for the x-axis except
    - Orientation: Allows you to angle the axis values for optimal placement.
  - Label: Same as for the x-axis
  - Scale: The range of values displayed on axis
    - Min/Max Value: Allows you to override the minimum/maximum value displayed on the axis.
    - Logarithmic: Used for graphs that have very large variances in magnitude.

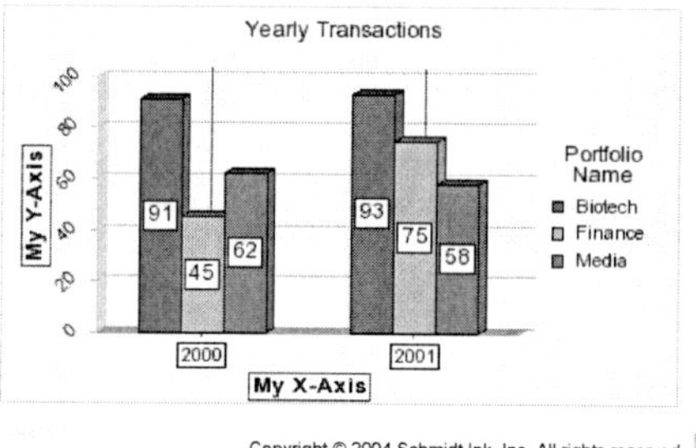

Slide 106

We may want to format the Y-axis for the same reasons we want to format the x-axis. However, there are two other reasons we may want to format the Y-axis – to adjust the scale or modify the value orientation.

Format the Y-Axis

1) Click on the Chart created in the previous example to select it.
2) Change the Values Orientation
   Click the cell to the right of the Orientation option in the Values section.
   Click the little down arrow button in the cell.
   Select 45.
3) Change the axis label
   Enter My Y-Axis into the Other Label option in the Label section.
4) Place a border around the label
   Place your cursor over the cell to the right of the Borders option in the Values section.
   When the small button appears, click on it.
   Select a style, color, and click the Box border button.
   Click Ok.

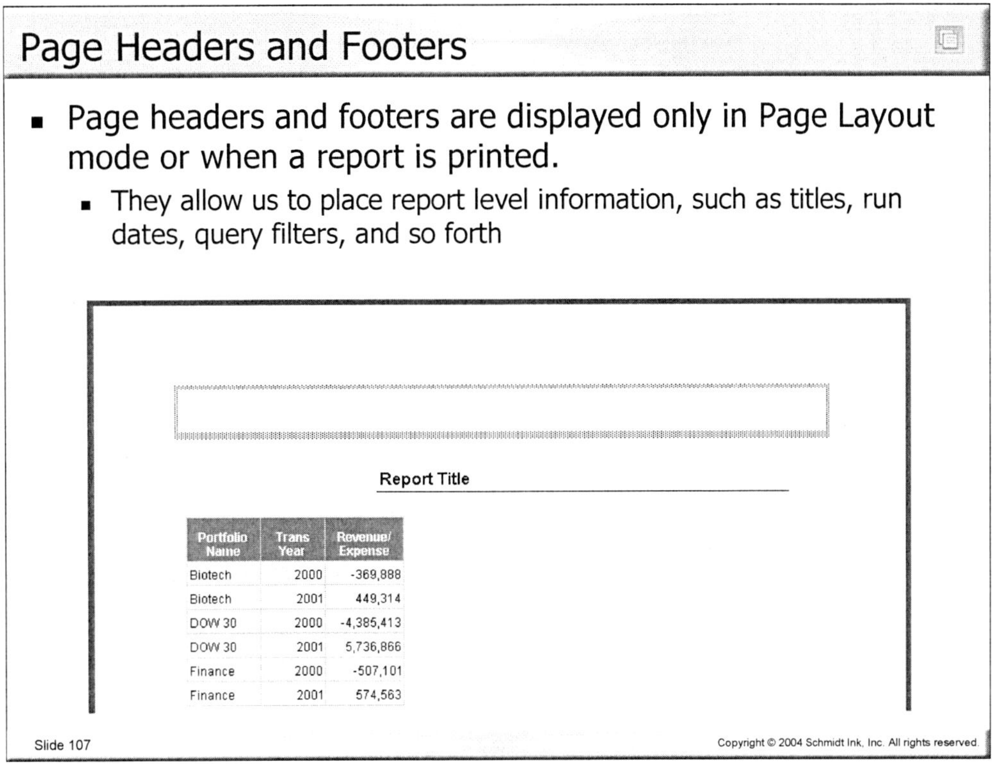

Page Headers and Footers
Page headers and footers are displayed only in Page Layout mode or when a report is printed.
They allow us to place report level information, such as titles, run dates, query filters, and so forth
Report Title
Portfolio Name
Trans Year
Revenue/ Expense
Biotech 2000 -369,888
Biotech 2001 449,314
DOW 30 2000 -4,385,413
DOW 30 2001 5,736,866
Finance 2000 -507,101
Finance 2001 574,563
Slide 107
Copyright © 2004 Schmidt Ink, Inc. All rights reserved.

Page headers allow us to place report level information, such as titles, run dates, criteria values, page numbers, and so forth. Almost every report that is to be printed or viewed in a page layout includes information in the page header and/or footer.

Using the Page Header and Footer

1) Create a document with Portfolio Name, Trans Year, Portfolio Company, Revenue/ Expense, and Num Transactions.
2) Set Portfolio Name as a Section.
3) Click on the white-space of the section to select it.
4) Click the Properties tab to activate the tab.
5) Expand the Page Layout section.
6) Check the Avoid Page Break in Section option.
7) Click the Page Layout option to expose the page header and footer.
8) Click on the header area to show the header outline (as seen in the slide).
9) Click the Templates tab to activate the tab.
10) Expand the Free-standing Cells section and drag the Blank Cell to the left potion of the header.
11) Click on the cell and enter My Report into the Formula toolbar.
12) Drag the Last Refresh Date cell to just beneath the title cell.
(Be sure that the cursor is in the header)
13) Scroll to the bottom of the page and click in the footer area to show the border of the footer.
14) Drag the Page Number/Total Pages cell into the footer.

Free-Standing Cells

- Free-standing cells inhabit almost every report in existence. They hold the titles, the page numbers, the query filters, measure summaries, and so forth.
- There are two ways to get a free-standing cell on to a report
  - Drag one of the Free-Standing Cell templates onto a report.
  - Drag a measure from the Data tab or from a report structure, such as a table, and drop it onto the report.

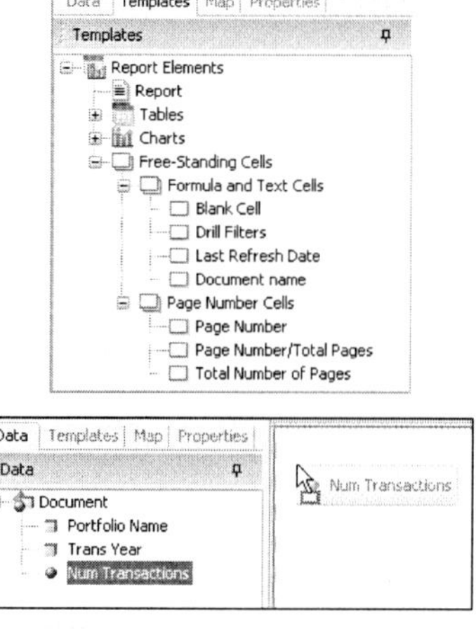

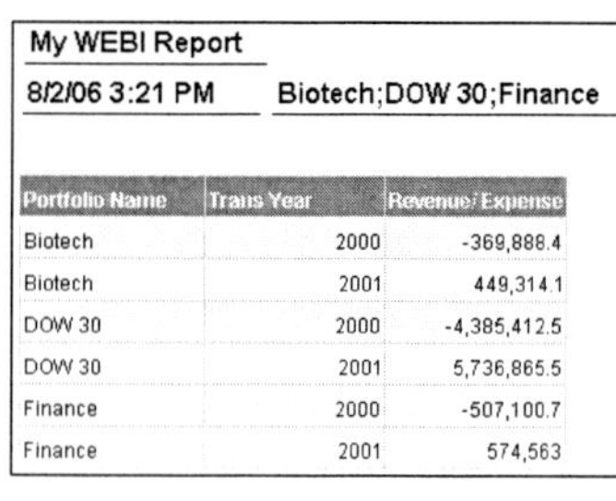

My WEBI Report

8/2/06 3:21 PM Biotech;DOW 30;Finance

Portfolio Name	Trans Year	Revenue/Expense
Biotech	2000	-369,888.4
Biotech	2001	449,314.1
DOW 30	2000	-4,385,412.5
DOW 30	2001	5,736,865.5
Finance	2000	-507,100.7
Finance	2001	574,563

Slide 108

Free-standing cells are used to reflect all sorts of report data. They can contain last refresh dates, page numbers, query filters, measure summaries, report titles, and the list goes on. Free-standing cells hold singular data. They cannot hold data that has more than one value, as a table can.

Create Some Free-Standing Cells

1) Create a report with Portfolio Name, Trans Year, and Revenue/Expense. Include the Portfolio Names query filter in the query.
2) Save the document as "My WEBI Report"
3) Right-click on the Report Title cell and select Remove from the pop-up menu.
4) Click on the Templates tab to activate it.
5) Drag the Document Name template unto the report, in the upper-left corner.
6) Drag the Last Refresh Date template to just beneath the Document Name cell. (You can size the cell by dragging the right-hand border, similar to sizing a column)
7) Click the View Page Layout button to display the report header and footer.
8) Drag the Page Number/Total Pages template into the footer.
9) Drag the Blank Cell template and drop it just to the right of the Last Refresh Date.
10) Right-click on the cell and select Formula Toolbar from the pop-up menu. (The formula toolbar will appear in the toolbar area of Web Intelligence)
11) Enter the following into the edit field of the toolbar (It must be typed exactly as it appears)
    =UserResponse("Please select portfolio names:")
    We will learn more about formulas later in this class. So, if this step is difficult, then ignore it till we can cover it in more detail.

Free-Standing Cells (Dragged from Report Structure)

- When you drag a dimension from the Manager or from a report structure, the report will turn into a Section report with the dropped cell as the master, as shown in this example, where Alternative Energy is the section value.
  - If a dimension or detail is put in a free-standing cell, and there is more than one value for the object, then the free-standing cell will display the #Error message.
- If a measure is dropped unto the report, it will simply aggregate to the report total, as shown in this example.
  - It will aggregate before any other formula that may be applied to it.
    For example, = Average ([Num Transactions]) in a freestanding cell will be the total number of transactions, not the average number of transactions.

Alternative Energy

214,664.6

Trans Year	Trans Quarter	Revenue/ Expense
2000	3	642
2000	4	-89,863
2001	1	14,341
2001	2	228,017
2001	3	61,529

Slide 109

We place free-standing cells on reports for a variety of reasons. We use them to display the value of the section master in section reports, which is usually a dimension dragged from a table on the report. We used them to display measure summaries, such as total sells or number of clients.

When a measure is placed in a free-standing cell, it will calculate its default calculation before any other formula or function that may be applied to it. For example, the formula, = Average ([Num Transactions]), will sum before it averages. The average of any sum is simply the sum itself, since the sum only represents one value. In the next chapter, we will learn how to create formulas that allow us to tell the formula to average across a context defined by dimension objects in a report. This formula will allow us to determine calculations in a free-standing cell, such as average yearly revenue, average portfolio number of transactions, and so forth.

Dragging Cells from a Report Structure

1) Create a report with Portfolio Name, Trans Year, and Revenue/Expense.
2) Drag Portfolio Name from the table and drop it above the table.
   High enough, so as not to create a crosstab
   This will create a section report with Portfolio Name as the master.
3) Move the table a little lower on the report
   (Click on the edge of the table and drag it down about a free-standing cell's height.)
4) Drag the Revenue/Expense object from the table to the report, just under the portfolio name master cell.
   (Before you release the mouse button, hold down the [Ctrl] key to copy it.)

Free-Standing Cells (Properties)

- Free-standing cells have several very useful formatting properties, in addition to most of the standard ones
  - Read Cell Content As: We can decode the contents of the cell as, Text, Hyperlink, HTML, or Image URL. This allows us to get very creative with reports.
  - Background Image: We can select a skin or Image URL as the background for selected cells.
  - Number Format: Allows us to format numbers and dates in selected cells.
  - Repeat on Every Page: This option allows us to repeat a cell on every printed page. It also works in Page Layout mode. We often use this option in Section reports to repeat the Master Cells.
  - Sorts: Allows us to sort the sections in a Section report.

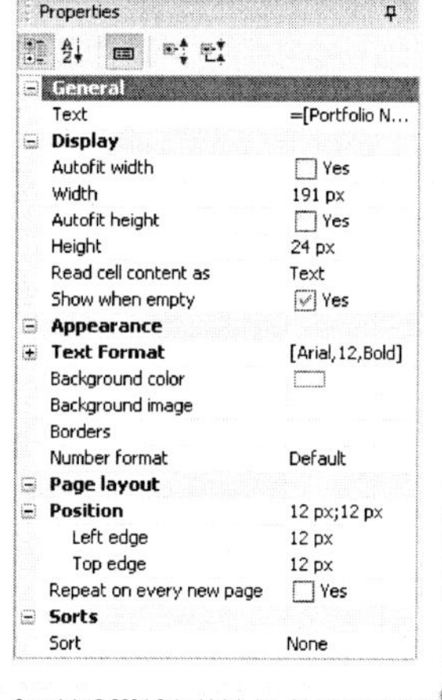

Slide 110

As with most report structures, if we use them, then we probably want to format them. I can't think of one professional report that I ever made where I did not have to format something. Knowing some of the special formats makes report design much easier. I hope that the few in this slide will help you.

You can add background skins to the system by copying *.gif files into the Business Objects Enterprise 11.5\Images\ directory on the Web Server. Some companies put their logo in this directory for convenient insertion into reports.

Create Some Free-Standing Cells

1) Using the report from the previous exercise...
2) Click on the section master cell to select it.
3) Click on the Properties tab to activate it.
4) Place your cursor over the cell to the right of the Background Image option, click the little button when it appears.
5) Select the Skin option and select a skin from the drop-down list.
6) Click Ok.
7) Check *Yes* for the *Show on Every New Page* option.
8) Click the cell to the right of the *Sort* option and select Descending.

Hyperlinks in Cells

- We can place hyperlinks in Cells that will allow us to
  - Jump to a Web site, as in http://www.SchmidtInk.com
    or
    <a href="http://www.SchmidtInk.com">Schmidt Ink Web Site</a>
  - Initiate an email, as in mailto://rschmidt@schmidtink.com
    or
    <a href="mailto:rschmidt@SchmidtInk.com
    ?subject=Great Course">Email Schmidt</a>
- To get Web Intelligence to treat the contents of a cell as a hyperlink, simply
  - Click on the Properties tab and set the *Read Cell Content As* option to hyperlink.

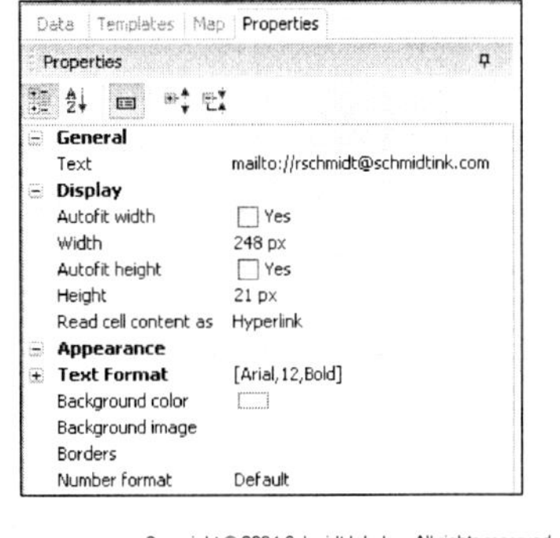

Slide 111

Sometimes we want to place a hyperlink in a cell to allow online viewers of the report to jump to a Web site or to send an email. The code for these hyperlinks can be typed into a cell, or can be supplied to each row by a data provider (query).

Create a Report with Hyperlinks

1) Create a report with Portfolio Name.
2) Click on the Templates tab of the Manager.
3) Expand the Free-standing Cells folder.
4) Expand the Formula and Text Cells folder.
5) Drag four Blank Cells onto the report.
6) Enter the following into one of the cells
   http://SchmidtInk.com
7) Enter the following into one of the other cells
   <a href="http://www.SchmidtInk.com">Schmidt Ink Web Site</a>
8) Enter the following into one of the last of the cells
   mailto://rschmidt@schmidtink.com
9) Enter the following into the last of the cells
   <a href="mailto:rschmidt@SchmidtInk.com?subject=Great Course">Email Schmidt</a>
10) Click on one of the cells to select it. Then while holding down the [CTRL] key, select the other three.
11) Click on the Properties tab of the Manager and set the Read Cell Content As option to hyperlink.

Working with Report Structures Summary

- In this chapter
  - We examined the report structures that we can place in reports.
  - We also examined different types of report formats, such as Sections and Breaks.
  - We looked at the Properties for each report structure and discussed how to use them in several examples.

Slide 112

This has been a very important chapter, because it introduced us to the many report structures and formats that we can use in Web intelligence. With experience you will become more adept at selecting which ones are best for the reports that you create for your company.

Schmidt Ink, Inc.

Creating Documents with Web Intelligence XI

Formulas, Variables
and
Various Functions
(Java Report Panel)

Slide 113

Introduction

- In this chapter, we will cover formulas and variables, and how to use them to create desired results in our documents.
- Now that we know how to create queries and report structures, it is a good time to talk about what we can do with this knowledge. Most reports contain formulas that perform analysis on our data to make it more understandable. In this chapter...
  - We will learn about formula syntax
  - We will learn how to create a variable
  - We are going to look at each of the function groups and create sample reports with functions from many of the groups
  - We will look at the If-Then-Else logic structure

Slide 114

Formula Syntax

- A formula in Web Intelligence is any expression that begins with an equal sign (=).
  - = Hi, = 5, = [Num Transactions] + 5, = "Hi"
  - All of the above is a valid formula, except the first (= Hi).
- Syntax is a formula's structure
  - All formulas must begin with an equal sign (=)
  - All text must be enclosed in double quotes
  - All objects and variables must be enclosed in square-brackets.
    - = [Num Transactions]
  - If there is more than one object in a context operator list, then they must be separated by semicolons and be enclosed in parentheses
    - = [Num Transactions] In ([Trans Year]; [Portfolio Name])

Slide 115

Formula syntax is important, because BusinessObjects will not accept a formula, unless it has proper syntax. The rules are simple and the Help file often has examples of proper syntax for the functions.

Some formulas are very complicated and it is hard for most people to understand the syntax. However, most formulas that are needed in reports are not that complex and with a little practice, the proper syntax can be easily achieved.

Formula syntax also follows the basic order of operations learned for math. Do what is inside the parentheses first, then multiply, then divide, then add, then subtract.

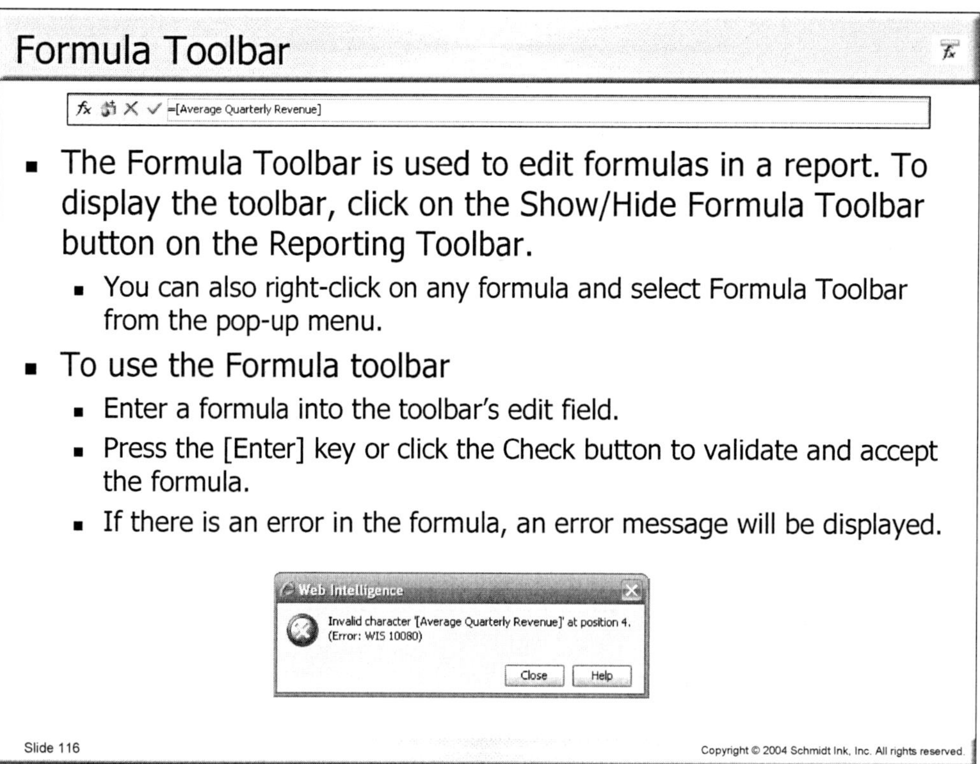

In Web Intelligence, we do not edit formulas directly in their cells on a report. We use the Formula Toolbar to edit the formulas. Once a formula is entered, then it can be validated and accepted by clicking on the Check button on the toolbar. If the formula is to be discarded, then just click the X button on the toolbar.

Formula Toolbar Function Tooltips

=MonthsBetween([Call Date];ToDate("1/1/2006"; "m/m/yyyy"))

date ToDate (string input_string; string date_format)

Applies a date format, specified by the string format, to a character string.

- Many reports contain functions that you may not be familiar with. If the formula for a selected cell is displayed in the Formula Toolbar, then you can see a short description by placing the cursor over the function in the toolbar.
  - In this example, the cursor is placed over the ToDate function. The tooltip displays the proper syntax for the function and also a short description.
  - This tooltip can also be very helpful when creating formulas in the toolbar that use functions.

Slide 117

Web Intelligence thoughtfully displays a tooltip when the cursor is placed over a function in the toolbar.

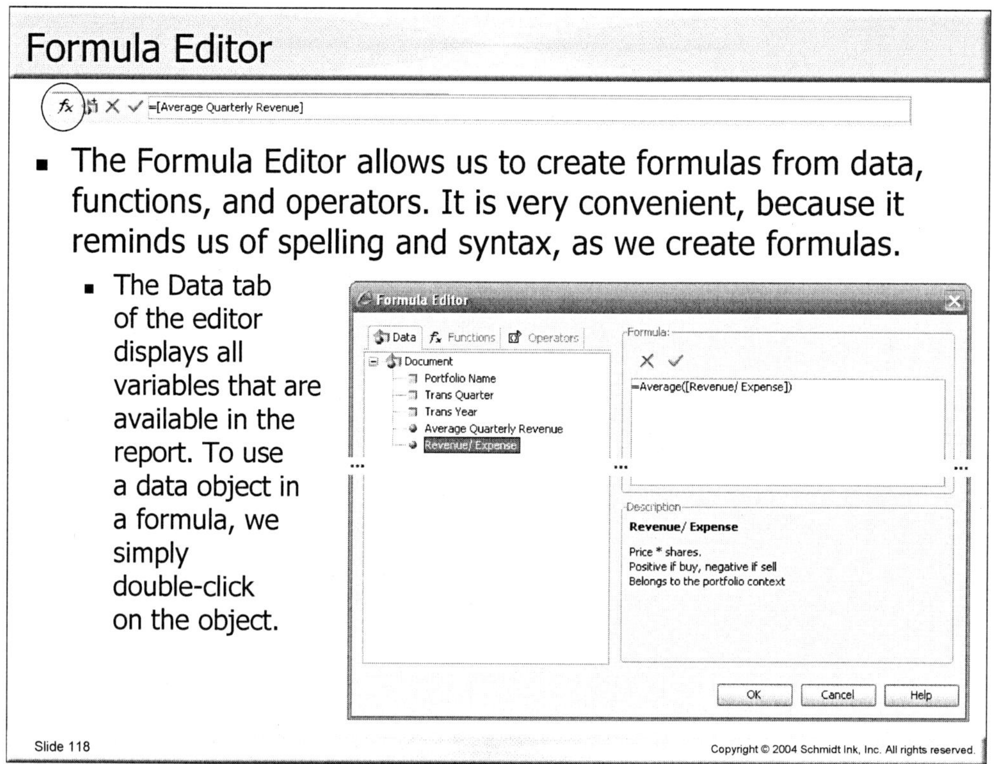
Formula Editor
=[Average Quarterly Revenue]
The Formula Editor allows us to create formulas from data, functions, and operators. It is very convenient, because it reminds us of spelling and syntax, as we create formulas.
The Data tab of the editor displays all variables that are available in the report. To use a data object in a formula, we simply double-click on the object.
Formula Editor
Data
Functions
Operators
Document
Portfolio Name
Trans Quarter
Trans Year
Average Quarterly Revenue
Revenue/ Expense
Formula:
=Average([Revenue/ Expense])
Description
Revenue/ Expense
Price * shares.
Positive if buy, negative if sell
Belongs to the portfolio context
OK
Cancel
Help
Slide 118
Copyright © 2004 Schmidt Ink, Inc. All rights reserved.

In most cases, it is much simpler to use the Formula Editor to create formulas. It is simpler, because instead of typing formulas, we just select data object names, functions, and operators from a list. This helps us not to misspell object names and to use the correct syntax. We can also validate complicated formulas at intermediate stages of completion. Another advantage is that we can see all of the available options, even ones that we may not have considered.

When a data object (also known as a variable) is selected, the Description section of the dialog will display any description (metadata) that the universe designer has assigned to the object. All objects have descriptions in some universes, and in other universes none of them may have a description. In most universes at least a few objects have a description, especially the ones that are confusing.

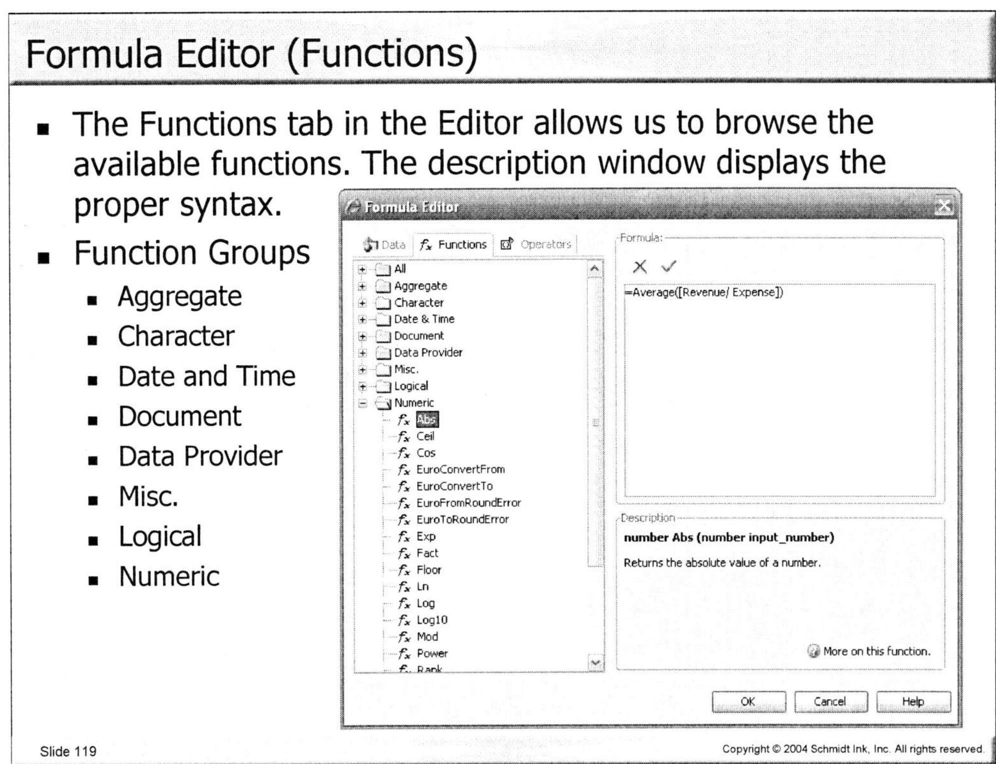

Web Intelligence has many powerful and useful functions. Almost every report created uses at least one function, such as LastRefreshDate, Page, DocumentName, and many more. Web Intelligence has categorized the functions into groups to make them easier to locate. In this chapter, we will use many of the functions from each group.

More on Functions

- Functions return information or a result to a formula
  - = Sum ([x]), = CurrentDate(), = ToNumber("5"), ...
  - Functions can return a date, text, or a number.
- Functions can accept arguments
  - An argument is the data within the parentheses that supply information to the function.
  - Arguments can be dates, text, numbers, or other functions
- The description for a selected function is displayed in the Description area of the Formula Editor.
  - Usually this description is very simple and used only as a reminder of the proper syntax.
  - The More on this function hyperlink will display much more in depth information on a selected function.

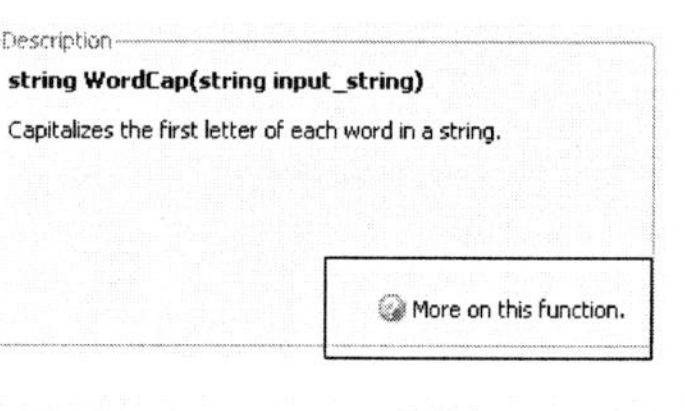

Slide 120

Functions may or may not accept an argument. An argument is a value or object that the function is to work on. For example, the Sum function takes one argument and it is a data object (also known as a variable). It will then aggregate this object's values across a defined context, such as a column in a table.

Most arguments are type sensitive. This means that if the function expects a date for an argument, then you cannot supply a string. For example, the formula

= MonthsBetween ([Trans Date], "1/1/2006")

will not work, because the second argument is of the wrong type. The MonthsBetween function expects two dates, not a date and a string (a string is also known as text).

All functions return a value. This is the purpose of functions – to return a value to a cell or formula. The values that functions return are also typed, meaning they return a number, a date, or a string (text).

Now that we know functions return typed values, we can fix the above formula using the ToDate function

=MonthsBetween ([Call Date]; **ToDate**("1/1/2006"; "m/m/yyyy"))

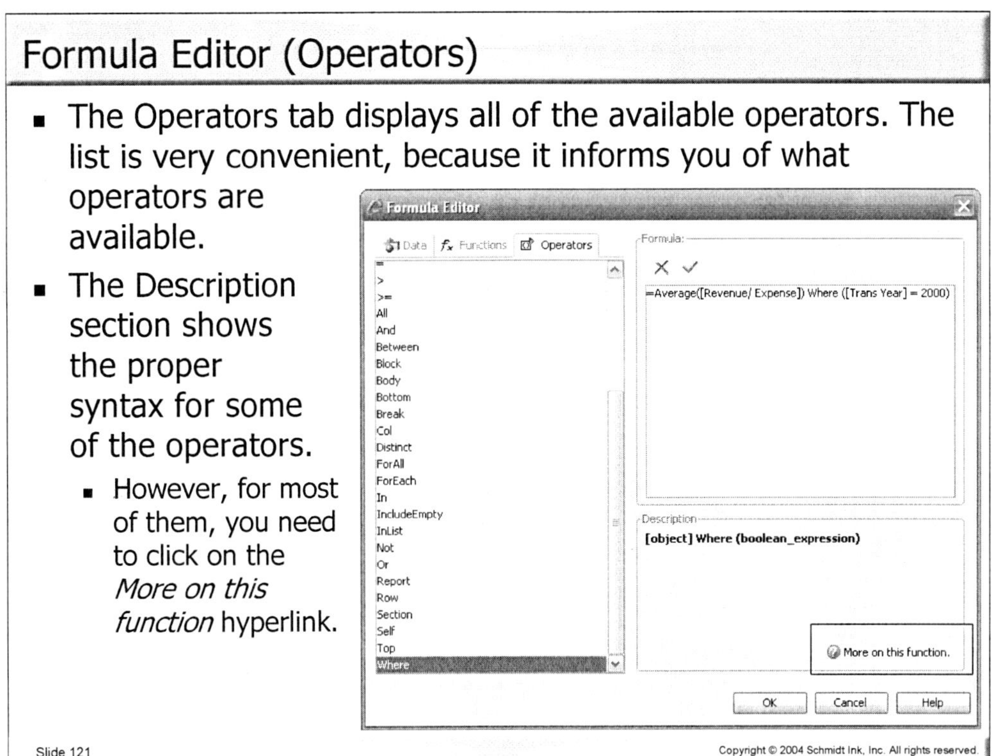
Formula Editor (Operators)
The Operators tab displays all of the available operators. The list is very convenient, because it informs you of what operators are available.
The Description section shows the proper syntax for some of the operators.
However, for most of them, you need to click on the *More on this function* hyperlink.
Formula Editor
Data
Functions
Operators
=
>
>=
All
And
Between
Block
Body
Bottom
Break
Col
Distinct
ForAll
ForEach
In
IncludeEmpty
InList
Not
Or
Report
Row
Section
Self
Top
Where
Formula:
=Average([Revenue/ Expense]) Where ([Trans Year] = 2000)
Description
[object] Where (boolean_expression)
More on this function.
OK
Cancel
Help
Slide 121
Copyright © 2004 Schmidt Ink, Inc. All rights reserved.

Web Intelligence includes many useful operators that allow us to make very powerful formulas. We will discuss many of them in the following pages.

More on Operators

- We use operators in formulas to enhance the functionality of the formula. Some operators simply instruct how to combine two statements, such as +, -, /, *, and so forth. Some work with functions to instruct them on how to handle data, such as Self, Top, Bottom, Row, and Col. And, others work with contexts within a report, such as ForEach, In, and Report.
  - We will not discuss the purpose of each operator here, but we cover them in context of their purpose. For example, we will discuss the context operators when we talk about that subject.
- It is necessary to master all of the operators to become a true Web Intelligence professional. They are essential to most powerful reports.

Slide 122

Since Business Objects works on sets of data, it has included many operators to help us instruct formulas and functions to achieve their purpose. In many cases they are necessary to the completion of a report. In this chapter and throughout most of the reminder of this book, we will use operators in most our formulas.

Variables

- Variables are simply packaged formulas with a name. They are created by packaging a formula in the Formula Toolbar.
- Variables are used in the same ways that objects from the data provider are. Once a variable is created, the distinction from a data provider object is not obvious.

=[Revenue/ Expense]

Create Variable

Variable Definition

Name: Revenue Container

Qualification: Measure

Formula: =[Revenue/ Expense]

OK Cancel Help

Data Templates Map Properties

Data

Document

Portfolio Name

Trans Year

Revenue Container

Revenue/ Expense

Arranged by: Alphabetic order

Slide 123

Variables are sometimes easier to use than formulas, because they are less bulky and can be represented like any other object in the data provider.

In this example, we are simply going to encapsulate a measure within a variable. This is often very useful, because after complicated reports are created, it is not uncommon to change the measure in the report. This usually means changing every formula in the report to accept the new measure. Not only is this a slow and frustrating process, it can also be very prone to errors.

Create a Holding Variable

1. Create a document with Portfolio Name, Trans Year, and Revenue/Expense
2. Select the data in the Revenue/Expense column by clicking on any of the values
3. Click on the Show/Hide Formula Toolbar to display the formula bar.
4. Click on the Create Variable button on the Formula Toolbar.
5. Name the Variable Revenue Container.
6. Keep the Measure Qualification.
7. Click Ok.

Finding and Editing Variables

- Since variables and objects from a query appear the same in the Data tab of the Manager, it is difficult to locate existing variables in a document. If you place your cursor over objects in the Data tab, a tooltip will reveal the variables by displaying the formula used to create it. Objects that are local to the query, display the description and not a formula.
- To edit an existing variable, right-click on it in the Manager and select Edit Variable.

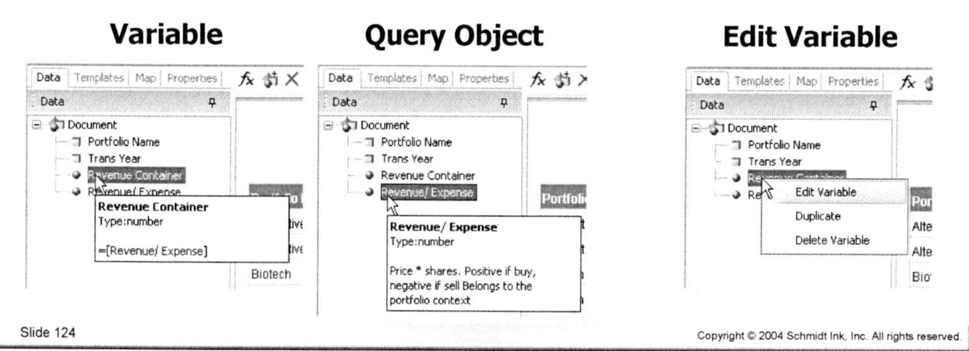

Documents can contain many variables. Many times, we need to view and edit the formula used to define a variable. To locate the local variables, place your cursor over a suspect variable in the Manager and view the tooltip. If the tooltip reveals a formula, then the object is a variable. To edit the variable, right-click on it and choose Edit Variable.

Edit the Holding Variable Created in the Previous Exercise

1. Edit the Query in the previous example and add Num Transactions.
2. Place your cursor over the Revenue Container and view the tool tip (=[Revenue/ Expense])
3. Right-click on the Revenue Container and select Edit Variable from the pop-up menu.
4. Change the name to Num Transactions Container.
5. Highlight the formula in the Formula edit and double-click on Num Transactions in the Data tab of the Editor.
6. Click Ok.
7. Click on the Revenue/ Expense header.
8. If the Formula toolbar is not visible, then click the Show/Hide Formula Toolbar button to display it.
9. Highlight the NameOf ([Revenue/ Expense]) formula and enter – Num Transactions.
10. Click the green check button to validate and accept the formula (Header Label).

IF Function

- The If function is probably one of the most important functions in BusinessObjects, because it allows us to add logic and act upon it in our reports.
- If function syntax
  - If (Boolean formula; True Expression; False Expression)
  - The Boolean formula is any logical statement that has a single outcome. For example, [X] = 5. When [X] = 5 the statement is true, else it is false.
  - The True expression gets evaluated if the Boolean expression is true. This can be a value, an object, or another formula (including another If function.)
  - The False expression gets evaluated if the Boolean statement is false.
  - Below is an example of an embedded If statement. The embedded If statement can populated either the True or False expressions, or both.
    If ([X]=5; [Revenue] *5; **If (**[X]=6; [Revenue]*7; 0**))**

Slide 125

The If function is one of the most important functions in Web Intelligence, because it allows us to add logic to our reports. With the If functions, we check values before we calculate, we substitute values, we calculate different expressions, and so forth.

Use the If Function to Check a Denominator Before Dividing

1) Create a report with Portfolio Name and Revenue/ Expense.
2) Add another query with Call Portfolio Name and Num Calls.
3) Click the Run Queries button.
4) Select the *Include the Result Objects in the Document without Generating a Table* option from the New Query dialog.
5) Merge the Portfolio Name and Call Portfolio Name dimensions.
    1) Click the Merge Dimensions button on the Reporting toolbar
    2) Select both Portfolio Name and Call Portfolio Name
    3) Click the Merge button and accept the default name.
    4) Click ok on all dialogs to return to the report.
6) Drag the Num Calls object to the right-side of the Revenue/ Expense column (This will create a new column that is populated with Num Calls)
7) Insert a column after Num Calls.
8) Highlight one of the new cells (A body cell in the new column)
9) Show the Formula toolbar and enter the following formula
   =If (IsNull ([Num Calls]); 0; [Revenue/ Expense]/[Nu.
   You can call this column Revenue per Call

Aggregate Functions

- Aggregate functions operate on all of an object's values in a context.
  - For example, sum([Num Transactions]) will sum the number of transactions in a column, in a break, in a section, ...
- Aggregate functions normally accept only numbers for arguments.
  - Count, Min, and Max are exceptions.
  - We often use the Min or Max function to rectify a #MULTIVALUE error

Portfolio Name	Trans Year	Num Transactions	Num Transactions
Alternative Energy	2000	78	= Sum([Num Transactions])
	2001	96	= Sum([Num Transactions])
Alternative Energy	**Sum:**	**174**	= Sum([Num Transactions])
Biotech	2000	91	= Sum([Num Transactions])
	2001	93	= Sum([Num Transactions])
Biotech	**Sum:**	**184**	= Sum([Num Transactions])
DOW 30	2000	675	= Sum([Num Transactions])
	2001	816	= Sum([Num Transactions])
DOW 30	**Sum:**	**1,491**	= Sum([Num Transactions])
Finance	2000	45	= Sum([Num Transactions])
	2001	75	= Sum([Num Transactions])
Finance	**Sum:**	**120**	= Sum([Num Transactions])
Media	2000	62	= Sum([Num Transactions])
	2001	58	= Sum([Num Transactions])
Media	**Sum:**	**120**	= Sum([Num Transactions])
Technology	2000	274	= Sum([Num Transactions])
	2001	317	= Sum([Num Transactions])
Technology	**Sum:**	**591**	= Sum([Num Transactions])
	Sum:	**2,680**	= Sum([Num Transactions])

Aggregate functions operate on sets of data. They sum, count, min, max, average, and find the variance. Notice that the formula is exactly the same for each row or context in a report, even though each row or context displays a different value for the formula. This happens, because each row, break footer, and report footer is defined by the dimension values of that context. For example, each row in the report is defined by Portfolio Name and Trans Year. This allows Web Intelligence to calculate a different sum for each unique combination of dimension values in the context. The first row is defined by
Energy and 2000. Web Intelligence calculates the sum of the values for
native Energy and Trans Year = 2000 and displays 78 for the
is defined only by the Portfolio Name dimension.
Intelligence can place the sum of 174.
e entire context). Therefore
in the table (or the table total).

n a single cell, such as a
s, because most dimensions
alue, there is no guarantee that
properly display a dimension
or that dimension in the report.

en it is reasonable to think that the
highest alpha value. However, if
simply return that value. Hence, if
eport, then the formula

cell.

Measures and Aggregate Functions

- Measures often have built-in aggregate functions.
  - Most often, it is the Sum function
  - The Universe designer assigns the aggregate to the measure when it is created.

Portfolio Name	Trans Year	Num Transactions	Num Transactions
Technology	2000	274	= [Num Transactions]
	2001	317	= [Num Transactions]
Technology	**Sum:**	**591**	= [Num Transactions]
	Sum:	**2,680**	= [Num Transactions]

- The built-in function will always execute before any applied functions.
  - For example, with the formula = Average ([Num Transactions]), the number of transactions will conform to the report context, before it is averaged. Until it simply degenerates to the sum function, because the average of a single sum is the sum.

Average	**2,680**

Portfolio Name	Num Transactions
Alternative Energy	174
Biotech	184
DOW 30	1,491
Finance	120
Media	120
Technology	591
Average	**447**

Portfolio Name	Trans Year	Num Transactions
Alternative Energy	2000	78
Alternative Energy	2001	96
Biotech	2000	91
Biotech	2001	93
DOW 30	2000	675
DOW 30	2001	816
Finance	2000	45
Finance	2001	75
Media	2000	62
Media	2001	58
Technology	2000	274
Technology	2001	317
	Average	**223**

Slide 127

Most measures have built-in aggregate functions that define how the measure will conform to different contexts within a report. As can be seen from the slide, the measures will conform to a context, before any external aggregate function is applied to the measure. This causes a weird anomaly, the average of a report can also be the sum.

To rectify this anomaly, BusinessObjects supplies us with context operators that will override the default report context for a measure. In this example, the ForEach operator will allow us to properly represent the averages in a freestanding cell. We will learn how to do this later in this class.

Running Aggregate Functions 3/4

- Running aggregates incrementally aggregate columns of data. They allow you to see an aggregate total as it evolves
- This table has two running sums in the last columns
  - Acc Comp sums the number of calls made by the entire company, thus it keeps summing through the break
  - Acc Man sums only for each manager, thus it resets with each new manager
- The Acc Man column uses a reset argument to reset the aggregate with each new manager.
- The two basic formulas are the following
  - = RunningSum([Num Calls])
  - = RunningSum([Num Calls]; ([Call Port Mgr Name]))

Call Port Mgr Name	Call Year	Call Quarter	Num Calls	Acc Man	% Man Calls	% Company Calls	Acc Company	% Total Calls
Robert Denning	2000	3	6	6	38%	8%	41	52%
	2000	4	2	8	50%	10%	43	54%
	2001	1	5	13	81%	16%	48	61%
	2001	2	3	16	100%	20%	51	65%
Robert Denning			16	16	100%	20%	51	65%

Call Port Mgr Name	Call Year	Call Quarter	Num Calls	Acc Man	% Man Calls	% Company Calls	Acc Company	% Total Calls
Sean Wilkenson	2000	3	13	13	46%	16%	64	81%
	2000	4	5	18	64%	23%	69	87%
	2001	1	6	24	86%	30%	75	95%
	2001	2	4	28	100%	35%	79	100%
Sean Wilkenson			28	28	100%	35%	79	100%
			79	79		100%	79	100%

Slide 128

Create a Report with Running Sums

1. Create a document with *Call Port Mgr Name*, *Call Year*, *Call Quarter*, and *Num Calls*
2. Insert five columns to the right of the Num Calls column
3. Insert a break on *Call Port Mgr Name*
4. Enter the following formula into Acc Man

   = RunningSum ([Num Calls]; ([Call Port Mgr Name]))
5. Enter the following formula into % Man Calls

   = RunningSum ([Num Calls]; ([Call Port Mgr Name])) /
   [Num Calls] In ([Call Port Mgr Name])
6. Enter the following formula into % Company Calls

   = RunningSum ([Num Calls]; ([Call Port Mgr Name])) / [Num Calls] In Report
7. Enter the following formula into Acc Company

   = RunningSum ([Num Calls])
8. Enter the following formula into % Total Calls

   = RunningSum ([Num Calls]) / [Num Calls] In Report
9. Copy each formula down into the break footer.
10. Select the table by clicking on the outer halo and activate the Properties tab of the Manager.
11. Check the *Show Footers* option.
12. Copy the formulas into the table footer.

To make the table look exactly like the one in the slide, you can insert a column to the right of % Comp Calls and format it with no shading and no borders.

Aggregate Functions - Reset Operators (Col, Row)

- In the previous example, we saw how the reset dimension caused the RunningSum function to reset to zero at each new value (Break) of the dimension.
- In a crosstab, we can use the keyword Row or Col to reset the aggregate function. This keyword also works with the Percentage function.

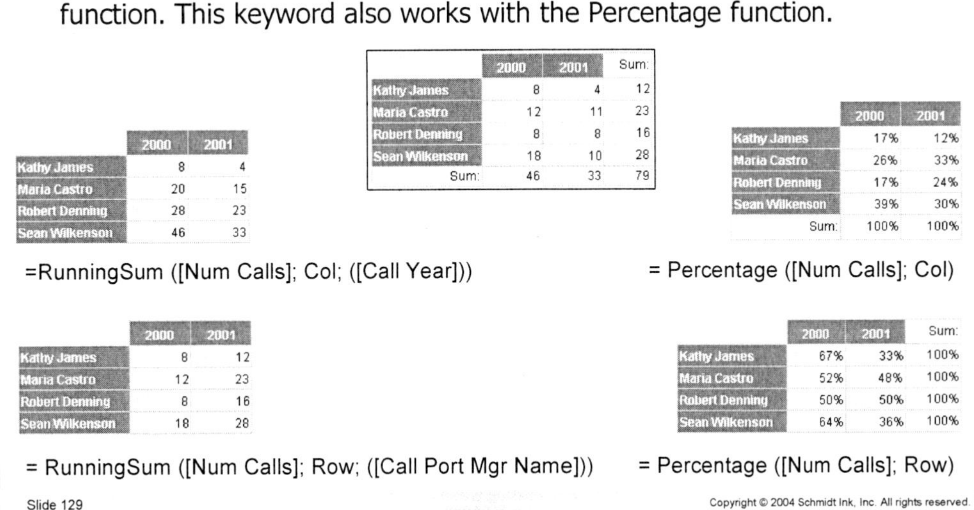

	2000	2001	Sum:
Kathy James	8	4	12
Maria Castro	12	11	23
Robert Denning	8	8	16
Sean Wilkenson	18	10	28
Sum:	46	33	79

	2000	2001
Kathy James	8	4
Maria Castro	20	15
Robert Denning	28	23
Sean Wilkenson	46	33

=RunningSum ([Num Calls]; Col; ([Call Year]))

	2000	2001
Kathy James	17%	12%
Maria Castro	26%	33%
Robert Denning	17%	24%
Sean Wilkenson	39%	30%
Sum:	100%	100%

= Percentage ([Num Calls]; Col)

	2000	2001
Kathy James	8	12
Maria Castro	12	23
Robert Denning	8	16
Sean Wilkenson	18	28

= RunningSum ([Num Calls]; Row; ([Call Port Mgr Name]))

	2000	2001	Sum:
Kathy James	67%	33%	100%
Maria Castro	52%	48%	100%
Robert Denning	50%	50%	100%
Sean Wilkenson	64%	36%	100%

= Percentage ([Num Calls]; Row)

Slide 129

The Running aggregates and the Percentage function both work with reset arguments. They need these arguments to tell them to reset and start calculating from zero. In crosstabs, the Row or Col keyword tells the functions to aggregate the rows or the columns. These keywords are necessary, because the body of a crosstab is both vertically and horizontally orientated. In the percentage function, the Row and Col operators act as both directional and reset arguments. For example, the keyword Col will cause the percentage function to operate individually on each column in a crosstab. However, the Running aggregates need both the Col or Row operator and a reset argument, as shown in the slide.

Create a Report with the Reset Operators (Col, Row)

1. Create a document with *Call Port Mgr Name*, *Call Year*, and *Num Calls*
2. Turn the table into a crosstab with Call Year in the Column header.
3. Copy the crosstab three times, so that there are now four.
4. Insert the following formula into one of the crosstabs
   =RunningSum ([Num Calls]; Col; ([Call Year]))
   Col - will cause the RunningSum function to calculate vertically down the columns.
   ([Call Year]) – will cause the function to reset with each new year (Column).
5. Insert the following formula into one of the crosstabs
   = RunningSum ([Num Calls]; Row; ([Call Port Mgr Name]))
   Row - will cause the RunningSum function to calculate horizontally across the rows.
   ([Call Port Mgr Name]) – will cause the function to reset with each manager (Row).
6. Insert the following formula into one of the crosstabs
   = Percentage ([Num Calls]; Col)
   Col - will cause the Percentage function to calculate vertically and reset with each new column.
7. Insert the following formula into one of the crosstabs
   = Percentage ([Num Calls]; Row)
   Row - will cause the Percentage function to calculate horizontally and reset with each new row.

Aggregate Functions – Percentage Function

- The Percentage function calculates the percent distribution of the measure values in a column.
  - Row Value / (Total of Column)
- The percentage function takes one measure argument
  - = Percentage ([Revenue/ Expense])
- The function will adjust to any context in which it is placed. For example, if we added Trans Year to the table, the Percentage function will recalculate to display the correct values.

Alternative Energy

Portfolio Company	Num Transactions	% Transactions	Revenue/ Expense	% Revenue
Active Power, Inc.	16	9.2%	19,647.8	9.2%
AstroPower, Inc.	23	13.2%	3,620.4	1.7%
Ballard Power Systems Inc.	24	13.8%	14,997.7	7.0%
Capstone Turbine Corporation	27	15.5%	52,854.7	24.6%
Electric Fuel Corporation	19	10.9%	-2,507.2	-1.2%
FuelCell Energy, Inc.	24	13.8%	90,574.9	42.2%
H Power Corp.	19	10.9%	7,438.4	3.5%
Plug Power Inc.	22	12.6%	28,037.9	13.1%
Sum:	174	100.0%	214,664.6	100.0%

Slide 130

The percentage function is very powerful, because it allows us to display percent distributions in a relatively easy to understand manner. We used to have to use a function that included context operators, (we still do if we want percents over different contexts), now we can simply use the Percentage function.

Create a Report with the Percentage Function

1. Create a document with *Portfolio Name, Portfolio Company, Trans Year, Num Transactions, and Revenue/Expense*. Also add the *Portfolio Names* query filter.
   Only chose one Portfolio Name when you run the query.
2. Remove the Portfolio Name and Trans Year columns.
3. Drag Portfolio Name object from the Manager and drop it on the title free-standing cell.
   (Notice it will not display #MULTIVALUE, because there is only one Portfolio Name in the report.)
4. Insert a column before and after Revenue\ Expense in the table.
5. Insert the following formula into the column preceding Revenue\ Expense
   = Percentage ([Num Transactions])
6. Insert the following formula into the column following Revenue\ Expense
   = Percentage ([Revenue/ Expense])

Advanced Steps

1. Drag Trans Year from the Manager and drop it between Portfolio Company and Num Transactions. This will insert a column between with the Portfolio Manager and Num Transactions.
2. Place a break on the Portfolio Company
3. Copy all the calculated formulas to the break footer.
   (Click on any value in the column, drag it to the footer, hold down the [CTRL] key before releasing the mouse button)

Portfolio Name	Portfolio Company	Trans Year	Num Transactions	% Transactions	Revenue/ Expense	% Revenue
Alternative Energy	Active Power, Inc.	2000	6	37.5%	-17,569	-89.4%
Alternative Energy		2001	10	62.5%	37,217	189.4%
	Active Power, Inc.		16	9.20%	19,648	9.15%

Numeric Functions

- Numeric functions operate only on the current number in a context. They do not aggregate, as the aggregate functions.
- Numeric functions consist of rounding and mathematical functions, with a few exceptions
  - Rank, this function accepts two arguments – a dimension and a measure. It then ranks the dimension based on the aggregation of the measure.
    - Rank ([Revenue/Expense]; [Portfolio Name])
  - ToNumber, this function converts a text representation of a number to a numeric number.
    - This is often used in If-then statements when comparing text numbers to numeric numbers.
      - If (ToNumber ([TextNumber]) = 5; [X]; [Y])

Slide 131

Numeric functions allow us to create powerful analytical formulas in our reports.

Numeric Functions – ToNumber (MTD Report)

- While the ToNumber function may seem simple, it can allow us to create very efficient reports.
  - For example, the following report uses the ToNumber function to allow us to create a Whole Year – Month to Date Revenue report (With a single query).

Alternative Energy

Portfolio Company	WY 2000 Revenue	MTD 2001 Revenue
Active Power, Inc.	-17,568.8	26,381.1
AstroPower, Inc.	2,809.4	11,274.7
Ballard Power Systems Inc.	-16,380.3	-22,478.4
Capstone Turbine Corporation	7,634.5	27,606.7
Electric Fuel Corporation	-9,571	4,803.3
FuelCell Energy, Inc.	9,342.8	-61,580.7
H Power Corp.	-24,469	-7,494.1
Plug Power Inc.	-41,018.9	35,828.3
Sum:	-89,221.3	14,340.9

Slide 132

Create a Whole Year versus Month-to-Date Report

1. Create a report with Portfolio Name, Portfolio Company, Trans Year, Trans Month, and Revenue/ Expense. Use the following condition.
   - Drag the Trans Month dimension to the Query Filters window.
   - Select the *Less Than Or Equal To* operator
   - Select the *Prompt* operand and enter - Enter Current Month Number
   - Drag Trans Month to the Query Filters window (Again, now there are two)
   - Select the *Greater Than* operator
   - Select the *Prompt* operand and enter - Enter Current Month Number (You should copy it from the first).
2. Change the And operator to an Or operator, if necessary.
3. Click Run Query.
4. Enter 3 in the prompt dialog.
5. Delete the Trans Year, Trans Month, and Revenue/ Expense columns.
6. Insert two columns to the right of the Portfolio Company column.
7. Click on an empty cell in the new column.
8. Create the following two variables, by entering the formula into the Formula toolbar, then clicking the *Create Variable* button on the toolbar
   - Name: CurrentMonth
   - Formula: = ToNumber(UserResponse ("Enter Current Month Number"))
   - Name: IsMTD2001
   - Formula: = If ([Trans Year]=2001 And [Trans Month] <= [CurrentMonth]; 1; 0)
9. Delete any formulas in the new columns and enter the following headers and formulas
   - Header: WY 2000 Revenue
   - = [Revenue/ Expense] Where ([Trans Year]=2000)
   - Header: MTD 2001 Revenue
   - = [Revenue/ Expense] Where ([IsMTD2001]=1)
10. Make Portfolio Name a section master.
11. Place sums on the revenue columns.

Numeric Functions (Rank) – Ranking Example

- The Rank function allows us to rank dimensions in our reports by some measure in the report.
  - In this example, we will rank the top company in each portfolio. We will then echo the top company in the footer and compare it to the total companies within each portfolio.
- Rank Function Syntax
  - Rank([measure]; [dimension(s)]; TOP|BOTTOM; [reset_dimension(s)])

Portfolio Name	Portfolio Company	Company Rank	Revenue/ Expense
Alternative Energy	Active Power, Inc.	4	19,648
	AstroPower, Inc.	7	3,620
	Ballard Power Systems Inc.	5	14,998
	Capstone Turbine Corporation	2	52,855
	Electric Fuel Corporation	8	-2,507
	FuelCell Energy, Inc.	1	90,575
	H Power Corp.	6	7,438
	Plug Power Inc.	3	28,038
Total Companies			214,665
#1 Company	FuelCell Energy, Inc.		86,250
Rest of Companies			128,415
% #1 Company			40%

Slide 133

Ranking Dimensions in a Report

1. Create a report with Portfolio Name, Portfolio Company, and Revenue/ Expense
2. Make Portfolio Name the section master
3. Click on any Revenue/Expense value and click the Insert Sum button
4. Click on the footer row and insert three rows below the current footer
5. Insert a column to the left of Revenue/Expense
6. Insert the following into the new column
   - Header: Company Rank
   - Formula: = Rank ([Revenue/ Expense])
7. Create the following variable
   - Name: CompanyRank
     Formula: = Rank ([Revenue/ Expense]; [Portfolio Company]; [Portfolio Name])
   - Qualification: Dimension
8. In the second footer row enter the following
   - #1 Company
   - = Max(If ([CompanyRank]= 1;[Portfolio Company]; "*"))
   - = Max (If ([CompanyRank]= 1; [Revenue/ Expense]; 0))
9. In the third footer row enter the following
   - Rest of Companies
   - = [Revenue/ Expense] - Max (If ([CompanyRank]= 1; [Revenue/ Expense]; 0))
10. In the forth row enter the following
    - % #1 Company
    - = Max (If ([CompanyRank]= 1; [Revenue/ Expense]; 0))/ [Revenue/ Expense]

Character Functions

- Character functions allow us to manipulate text in our reports. Many of these functions are used often, below are a few examples
  - To insert a return character into text
    - "First Line" + char(13) + "Second Line"
  - To remove return characters
    - Replace ([Text]; char(13); " ")
  - To concatenate a date to text
    - [Text] + " " + FormatDate ([Date]; "Mmm dd YYYY")
  - To concatenate a number to text
    - [Text] + " " + FormatNumber ([Number]; "00000")
  - To isolate first name
    - = Substr ([Portfolio Mgr Name]; 1; Pos ([Portfolio Mgr Name]; " ") -1)
- The examples are almost endless and increase as the skill of the use of these functions increase.

Slide 134

Format Functions vs. Format Cell

- The Format functions, FormatDate and FormatNumber, convert dates and numbers to text.
  - Formatted data will only sort and behave like text
    - 01/01/2003
    - 01/01/2004
    - 01/02/2003
    - 01/02/2004
- If the data is to behave as it's native type, then it is better to use the Format->Cell dialog
  - Since characters represent the formats in the Format Cell dialog, a backslash must precede formatting characters to make them literals
    - Tra\de \Date: Mm/dd/yy
    - Num Transactions: 0

Slide 135

Formatting Cells

1. Create a document with *Portfolio Mgr Name*, *Trans Date*, and *Num Transactions*
2. Insert a column to the right of Trans Date and copy the Trans Date formula into the column.
3. Modify the first Trans Date column to the following formula

   = "Trade Date: " + FormatDate ([Trans Date] ; "mm/dd/yy")
4. Select the second Trans Date column
5. Right-click on any value in the column and select Format Number from the pop-up menu
6. Select the Date/Time category
7. Select the Custom check box and enter the following into the Data/Time format field

   \T\r\a\d\e \D\a\t\e: Mmm d, yyyy
8. Re-sort the table by placing a sort on the newly formatted column

The column with the FormatDate function will sort as text and not in chronological order. However, the formatted date column will still sort in chronological order, because the contents are still a date.

The backslashes are needed, because the 'd' character extracts the day number of the date. Without the backslashes the field may display

T12AM25238 DAMt238: 00/25/00

Date Functions

- Date functions allow us to manipulate dates in a document. Many of these functions are often used.
  - To display the day name of a date
    - = DayName ([Call Date])
  - To extract the day number of a month
    - = DayNumberOfMonth ([Call Date])
  - The extract the number of days between two dates
    - = DaysBetween ([Call Date]; CurrentDate())
  - To display the last day of the month for a date
    - = LastDayOfMonth ([Call Date])
  - To display the number of months between two dates
    - = MonthsBetween ([first_date]; [second_date])
  - To convert a text date to a date
    - = ToDate("1/1/2006"; "m/d/yyyy")
    - "m/d/yyyy" represents the format of the text date and may vary

Slide 136

Date Functions - Month End Report

- In this example, we will create a month end report. We will use many of the skills that we have learned in this chapter.
- The report shows the monthly revenue for each portfolio. It also displays the running sum of the revenue for each portfolio. The revenue is displayed in thousandths.

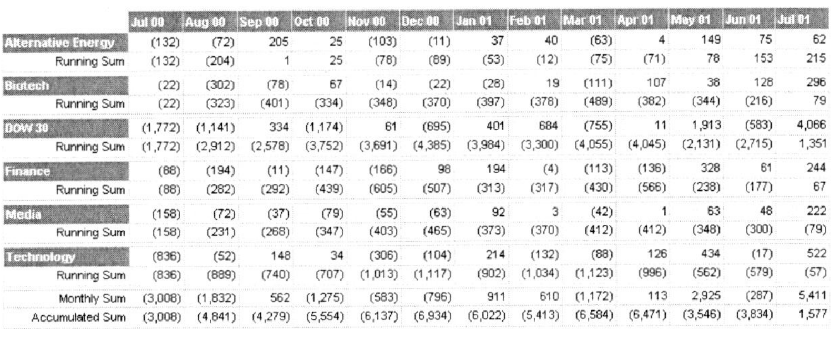

	Jul 00	Aug 00	Sep 00	Oct 00	Nov 00	Dec 00	Jan 01	Feb 01	Mar 01	Apr 01	May 01	Jun 01	Jul 01
Alternative Energy	(132)	(72)	205	25	(103)	(11)	37	40	(63)	4	149	75	62
Running Sum	(132)	(204)	1	25	(78)	(89)	(53)	(12)	(75)	(71)	78	153	215
Biotech	(22)	(302)	(78)	67	(14)	(22)	(28)	19	(111)	107	38	128	296
Running Sum	(22)	(323)	(401)	(334)	(348)	(370)	(397)	(378)	(489)	(382)	(344)	(216)	79
DOW 30	(1,772)	(1,141)	334	(1,174)	61	(695)	401	684	(755)	11	1,913	(583)	4,066
Running Sum	(1,772)	(2,912)	(2,578)	(3,752)	(3,691)	(4,385)	(3,984)	(3,300)	(4,055)	(4,045)	(2,131)	(2,715)	1,351
Finance	(88)	(194)	(11)	(147)	(166)	98	194	(4)	(113)	(136)	328	61	244
Running Sum	(88)	(282)	(292)	(439)	(605)	(507)	(313)	(317)	(430)	(566)	(238)	(177)	67
Media	(158)	(72)	(37)	(79)	(55)	(63)	92	3	(42)	1	63	48	222
Running Sum	(158)	(231)	(268)	(347)	(403)	(465)	(373)	(370)	(412)	(412)	(348)	(300)	(79)
Technology	(836)	(52)	148	34	(306)	(104)	214	(132)	(88)	126	434	(17)	522
Running Sum	(836)	(889)	(740)	(707)	(1,013)	(1,117)	(902)	(1,034)	(1,123)	(996)	(562)	(579)	(57)
Monthly Sum	(3,008)	(1,832)	562	(1,275)	(583)	(796)	911	610	(1,172)	113	2,925	(287)	5,411
Accumulated Sum	(3,008)	(4,841)	(4,279)	(5,554)	(6,137)	(6,934)	(6,022)	(5,413)	(6,584)	(6,471)	(3,546)	(3,834)	1,577

Slide 137

Create a Month End Report

1. Create a document with *Portfolio Name*, *Trans Date*, and *Revenue/Expense*
2. Create an xtab with the dates in the rows (column headers) position.
3. Place a break on the *Portfolio Name* column
4. Modify the Trans date row to the following

   = LastDayOfMonth ([Trans Date])
5. Format the Trans Date row using the Format Cell dialog

   Mmm YY
6. Modify the Revenue/Expense formula to the following

   =[Revenue/ Expense]/1000
7. Sum the Revenue, by clicking the Insert Sum button.
8. Modify the sum formula in the break footer, (-158) for Media, to

   =RunningSum ([Revenue/ Expense]; ([Portfolio Name])) /1000
9. Insert another report footer row and enter the following

   =RunningSum ([Revenue/ Expense]/1000; Col)
10. Delete the unnecessary right-edge sum column
11. Modify the headers

With this report you can ask, "Who was the first to show positive return?", "Who had positive cash the most months?", "Which portfolios had the most risk?"

Date Functions - Months since Last Call

- In this example, we use the MonthsBetween and the Fill function to create an in-line chart of how many months have past since a contact was last contacted.

Call Port Mgr Name	Call Contact Name	Last Call Date	Months Since Called	
Kathy James	Case, Steve	12/4/00	1	*
	Jeffrey Mallett	7/13/00	5	*****
	Kelly, Mike	7/10/00	5	*****
	Mel Karmazin	12/29/00	0	
	Michael D. Eisner	12/22/00	0	
	Richard J. Bressler	12/29/00	0	
	Robert A. Iger	12/22/00	0	
	Sumner M. Redstone	12/29/00	0	
	Terry S. Semel	12/19/00	0	
	Thomas O. Staggs	7/7/00	6	******

Slide 138

Create a Months Since Called Report

1. Create a document with *Call Port Mgr Name*, *Call Contact Name*, and *Call Date.* Use the following condition
   - Drag the Call Date object to the Conditions window
   - Select the Less Than or Equal to
   - Select the Prompt operand and enter: Enter Call Date
2. Run the report and enter 1/7/2001, into the prompt dialog.
3. Modify the *Call Date* header to *Last Call Date;* the column formula to

   = Max ([Call Date])
4. Insert a break on Call Port Mgr Name and delete the footer row for each break.
5. Create a variable

   Name: CurrentDate

   Formula: = ToDate (UserResponse ("Enter Call Date"); "m/d/yyyy")
6. Insert two columns to the right of *Last Call Date*
7. In the first column enter the following formula

   = MonthsBetween (Max ([Call Date]); [CurrentDate])
8. In the second column enter the following formula

   =Fill ("*"; MonthsBetween (Max ([Call Date]); [CurrentDate]))
9. Enter the following header

   Months Since Last Call

Logical Functions

- Logical functions return a true or a false. They are rarely used by themselves and almost always are used in the If function, filters, or hiding report elements
- Probably the two most common are
  - IsNull, is used to test if a value is empty
    If (IsNull ([variable]); 0; [variable])
    - This statement checks if a variable is NULL. If it is, then it replaces the NULL value with a zero. If it is not, then it uses the variable value.
  - IsError, is used to test if a variable is in error
    If (IsError ([variable]); 0; [variable])
    - This statement checks if a variable is in error, such as divide-by-zero. If it is in error, then it uses a zero. If it is not, then it uses the variable.

Slide 139

Use the IsNull function to Check a Denominator Before Dividing

1) Create a report with Portfolio Name and Revenue/ Expense.
2) Add another query with Call Portfolio Name and Num Calls.
3) Merge the Portfolio Name and Call Portfolio Name dimensions.
    1) Click the Merge Dimensions button on the Reporting toolbar
    2) Select both Portfolio Name and Call Portfolio Name
    3) Click the Merge button and accept the default name.
    4) Click ok on all dialogs to return to the report.
4) Drag the Num Calls object to the right-side of the Revenue/ Expense column and drop it
   (This will create a new column that is populated with Num Calls)
5) Insert a column after Num Calls.
6) Highlight one of the new cells (A body cell in the new column)
7) Show the Formula toolbar and enter the following formula
   =If (IsNull ([Num Calls]); 0; [Revenue/ Expense]/[Num Calls])
   You can call this column Revenue per Call

Document Functions

- Document functions return information about a document.
- The most commonly used are
  - Drill Filters, displays the drill path that has been taken in a report.
  - ReportFilter, displays any global filters that have been placed on a report. It will not display the filters that have been placed on individual blocks (tables, crosstab's, charts).
  - DocumentPartiallyRefreshed, this function is very important, because it will return true if a document is partially refreshed.
    - If (DocumentPartiallyRefreshed(); "Warning - Partially Refreshed!"; " ") Once a document is printed, it is very hard to determine if a document has been successfully refreshed. Therefore, it is a good idea to include this function somewhere on most of your reports.
  - Page and NumberOfPages, returns the current page number and the number of pages in a report.
    - Page() + " of " + NumberOfPages()

Slide 140

Use the ReportFilter Function

1) Create a report with Portfolio Name, Trans Year, and Revenue/ Expense
2) Click anywhere on the report white-space.
3) Click the Show/Hide Filter Pane button on the Reporting toolbar to show the Filter pane.
4) Drag the Portfolio Name dimension from the Manager into the Filter pane.
5) Select Equal to
6) Select List of Values and select Alternative Energy
7) Click on the report title to select it.
8) Click on the Show/Hide Formula Bar button on the Reporting toolbar
9) Enter the following formula
   = ReportFilter ([Portfolio Name])
10) Click on any value in the Revenue/ Expense column and click the Sum button
11) Insert a row under the sum and enter the following formula
   = [Revenue/ Expense] / NoFilter ([Revenue/ Expense])
   This formula will display the percent of the Filtered Portfolio Name of the entire report. At the time of this writing, this formula did not calculate properly, maybe it will be fixed in your version.

Data Provider Functions

- Data Provider functions return information about the data providers in a document.
- Almost every Data Provider function takes the name of a data provider has the first argument. Most data provider names are similar to *Query 1 with ....*
  - It is better to replace the name with the DataProvider function
    - DataProvider ([AnyObjectFromTheProvider])
- Examples of Data Provider functions are
  - LastExecutionDate (DataProvider ([obj]))
    - Can be used as Last Refresh Date or As Of Date
  - UserResponse ("Prompt Text")
    - Returns the user response to a prompt

Slide 141

Echoing Information on a Report

1. Create a document with *Portfolio Company*, *Trans Year*, *Trans Quarter*, *Num Transactions*, and *Revenue/Expense.* Place the *Portfolio Names* condition in the Conditions window. Click Run and select any portfolio.
2. Move the Report title to the left side of the report and enter

   = UserResponse ("Please select portfolio names:") + " Companies and Quarterly Report"
3. Insert a cell under the title cell and enter

   ="As of " + FormatDate (LastExecutionDate(DataProvider ([Portfolio Company])) ; "Mmm dd yyyy")
4. Place a break on Portfolio Company and place sums on Num Transactions and Revenue/Expense

Technology Companies and Quarterly Report
As of Oct 06 2004

Portfolio Company	Trans Year	Trans Quarter	Num Transactions	Revenue/ Expense
Advanced Micro Devices, Inc.	2000	3	1	-39,954
	2000	4	5	-16,494
	2001	1	2	22,365
	2001	2	3	22,533
	2001	3	1	4,228
Advanced Micro Devices, Inc.		Sum:	12	-7,322

Miscellaneous Functions

- Miscellaneous functions do not fit into any of the groups discussed so far.
- Examples are
  - NoFilter, this function removes the affects of filters on an object in a report.
  - Previous, allows a formula to reference the value of an object in the previous row of a report.

Trans Year	Trans Week	DOW 30 Transactions	Two-week Moving Avg	Acc DOW 30	% Total Tranactions
2000	2	113		113	53.81 %
2000	3	146	130	259	55.22 %
2000	4	145	146	404	55.12 %
2000	5	129	137	533	54.33 %
2000	6	142	136	675	55.10 %
2001	2	135	139	810	55.82 %
2001	3	176	156	986	56.38 %
2001	4	154	165	1,140	56.21 %
2001	5	147	151	1,287	56.08 %
2001	6	204	176	1,491	55.63 %

Slide 142

Two-Week Moving Average and Percent Total Report

1. Create a document with *Portfolio Name*, *Trans Year, Trans Week,* and *Num Transactions*
2. Place a filter on Portfolio Name equal to DOW 30
3. Delete the Portfolio Name column, but not the filter
4. Insert three columns to the right of Num Transactions
5. Enter the following into the first inserted column

   = If (Not (IsNull (Previous ([Num Transactions]))); ([Num Transactions] + Previous ([Num Transactions]))/2; "")
6. Enter the following into the next column

   = RunningSum([Num Transactions])
7. Enter the following into the last column

   = RunningSum ([Num Transactions]) / NoFilter (RunningSum ([Num Transactions]))

   This formula doesn't work well in my version
8. Modify the column headers as they are in the report in the slide

Opening Documents from a Hyperlink

- Web Intelligence allows us to open other documents in the Business Objects environment.
- We do this with the Open Document hyperlink.
  - ="<a href='http://boweb:8080/businessobjects/enterprise115/desktoplaunch/opendoc/openDocument.jsp?sDocName=Portfolio Details&lsSEnter+Portfolio+Name="+[Portfolio Name]+"&sWindow=Same&sRefresh=yes'>Please Click to get Portfolio Details</a>"
  - The syntax looks complicated, but it is simpler if we break it down...
    - (http://boweb:8080/businessobjects/enterprise115/desktoplaunch/opendoc/openDocument.jsp?) Is the address of your BO Server.
    - (sDocName=Portfolio Details) Is the name of the document.
    - (lsSEnter+Portfolio+Name="+[Portfolio Name]) Is the assignment of a prompt – Enter Portfolio Name
    - (&sRefresh=yes) Tells Web Intelligence to refresh the document.

Slide 143

Many times, we want to view other reports after viewing a current report. The report that we are viewing may cause us to want to see more or different information then the current report has.

In Web Intelligence, we can open other reports from within a document, by defining a hyperlink that uses the OpenDocument function. We use a formula that concatenates the needed pieces to form the correct hyperlink format. One syntax for the formula is displayed above. This syntax allows for an argument to be assigned a value from the current row in a report.

There is an example on the next page.

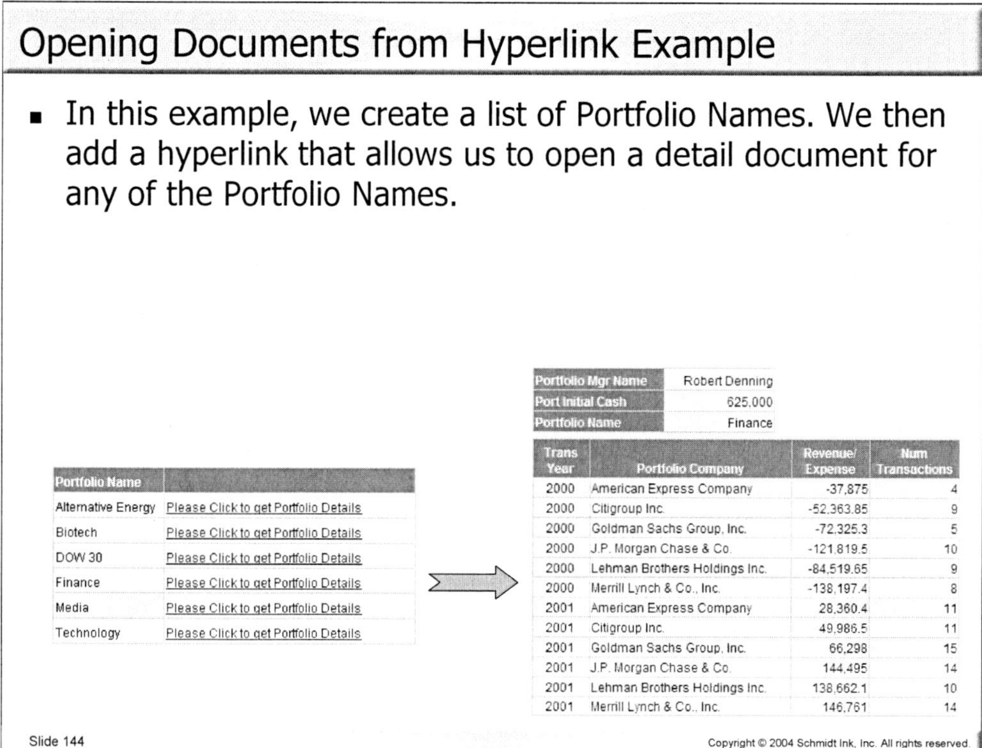

Portfolio Name	
Alternative Energy	Please Click to get Portfolio Details
Biotech	Please Click to get Portfolio Details
DOW 30	Please Click to get Portfolio Details
Finance	Please Click to get Portfolio Details
Media	Please Click to get Portfolio Details
Technology	Please Click to get Portfolio Details

Portfolio Mgr Name	Robert Denning
Port Initial Cash	625,000
Portfolio Name	Finance

Trans Year	Portfolio Company	Revenue/ Expense	Num Transactions
2000	American Express Company	-37,875	4
2000	Citigroup Inc.	-52,363.85	9
2000	Goldman Sachs Group, Inc.	-72,325.3	5
2000	J.P. Morgan Chase & Co.	-121,819.5	10
2000	Lehman Brothers Holdings Inc.	-84,519.65	9
2000	Merrill Lynch & Co., Inc.	-138,197.4	8
2001	American Express Company	28,360.4	11
2001	Citigroup Inc.	49,986.5	11
2001	Goldman Sachs Group, Inc.	66,298	15
2001	J.P. Morgan Chase & Co.	144,495	14
2001	Lehman Brothers Holdings Inc.	138,662.1	10
2001	Merrill Lynch & Co., Inc.	146,761	14

Create A Hyperlink Document Pair

1. Create a detail document with Portfolio Name, Port Initial Cash, Portfolio Mgr name. Trans Year, Portfolio Company, Revenue/ Expense, and Num Transactions.
2. Format the document into a desired format. You may watch the video to see how I did it.
3. Save the document as *Portfolio Detail.*
4. Create a new document with Portfolio Name.
5. Insert a column to the right of the Portfolio Name column.
6. Insert the following formula into the column.
   ="<a href='http://boweb:8080/businessobjects/enterprise115/desktoplaunch/opendoc/openDocument.jsp?sDocName=Portfolio Detail&lsSEnter+Portfolio+Name="+[Portfolio Name]+"&sWindow=Same&sRefresh=yes'>Please Click to get Portfolio Details</a>"
7. Click on any of the cells in the new column
8. Click on the Properties tab
9. Set the Read *Cell Content As* option to *Hyperlink.*
10. Click on any of the links to test out the formula.

For those of you with really good eyes, the formula is repeated here with very small font.
There is a space between Portfolio and Detail at the line break.
DocName=Portfolio Detail

="<a href='http://boweb:8080/businessobjects/enterprise115/desktoplaunch/opendoc/openDocument.jsp?sDocName=Portfolio Detail&lsSEnter+Portfolio+Name="+[Portfolio Name]+"&sWindow=Same&sRefresh=yes'>Please Click to get Portfolio Details</a>"

Formulas and Variables Summary

- This was a very important chapter, because we learned how to add valuable information to our reports through the use of formulas and variables.
- We learned
  - How to edit formulas in the Formula toolbar and in the Formula Editor.
  - How to use variables to make formula creation more efficient.
  - About the various functions that are available and we use a few to make some reports.

Slide 145

Formulas and Variables are very important to report creation. They allow us to diverge from simple listings of data to create powerful analytical reports.

Slide 146
Copyright © 2004 Schmidt Ink, Inc. All rights reserved.

Creating Documents with Web Intelligence XI

Data and Report Contexts (Java Report Panel)

Introduction

- In this chapter, we are going to discuss report and data contexts within a document.
- We are going to explore different ways to manipulate these contexts to help us make more complex reports.
- We will cover context operators
  - In, ForAll, and ForEach
  - Where
- Most complex reports manipulate default contexts to achieve desired output from formulas

Slide 148

Contexts in Reports

- Contexts in reports are created by structural breaks
  - Each row, header, break footer, section, and report header/footer are defined by different sets of dimension values
  - A sum in a row will aggregate for only that row; in a break footer, it will aggregate for all the rows in the break; in a report footer, it will aggregate for all of the rows in a report; and so forth

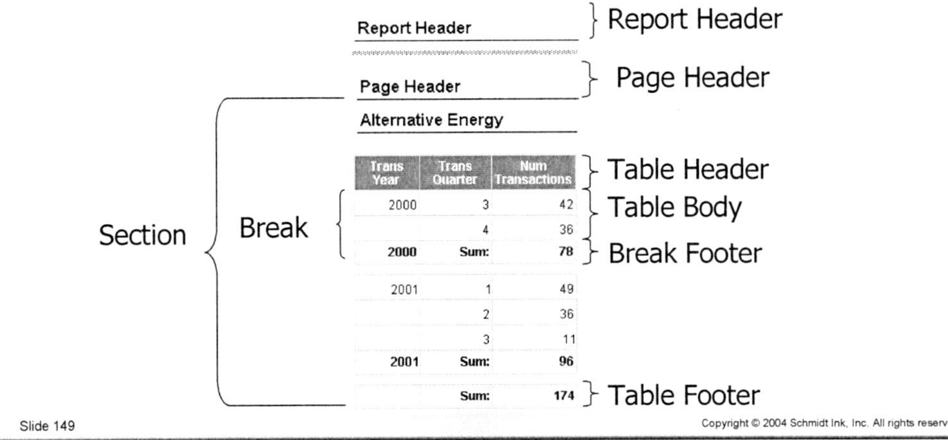

Slide 149

View Contexts in a Report

1. Create a report from the SI Equity universe.
   From the Portfolio class, select *Portfolio Name*, *Trans Year*, *Trans Quarter* and *Num Transactions*
2. Click Run Query
3. Set *Portfolio Name* as Section Master

   Right-click on a Portfolio Name and select Set as Section
4. Insert a break on *Trans Year*

   Select any value in the column and click the *Insert / Remove->Break* button on the Reporting toolbar
5. Place a sum on the *Num Transactions* column

   Click on any value in the column and click the *Inert Sum* button.

Notice that the sum is different in each of the contexts, even though the formula is the same for each context. The calculation engine in BusinessObjects sums the values based on the dimension values in the row.

Details values also affect the calculations. However, since there is only suppose to be one detail value per dimension value, it usually does not affect the calculation.

Contexts in Data

- Dimension values also form contexts in data
  - Even though the data is not in a structured report, the dimension values can still form contexts within the data.
  - In the tables below, Portfolio Name and Trans Year form contexts.
  - Additional contexts can be formed by combining the dimension values.
    - For example, Trans Year and Trans Quarter will form contexts:
      - 2000/3, 2000/4, 2001/1, 2001/2, 2001/3

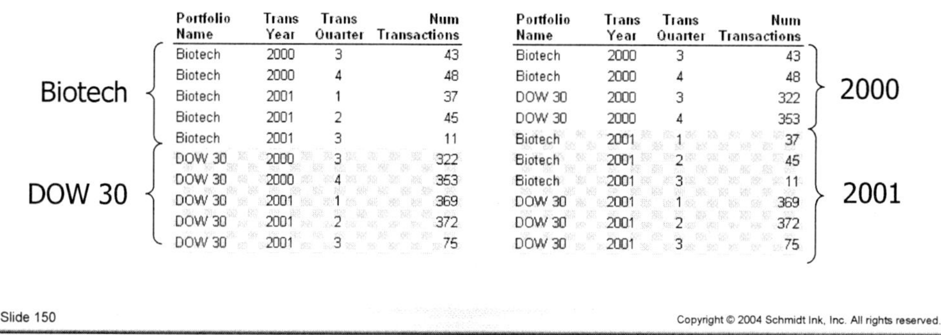

Portfolio Name	Trans Year	Trans Quarter	Num Transactions
Biotech	2000	3	43
Biotech	2000	4	48
Biotech	2001	1	37
Biotech	2001	2	45
Biotech	2001	3	11
DOW 30	2000	3	322
DOW 30	2000	4	353
DOW 30	2001	1	369
DOW 30	2001	2	372
DOW 30	2001	3	75

Portfolio Name	Trans Year	Trans Quarter	Num Transactions
Biotech	2000	3	43
Biotech	2000	4	48
DOW 30	2000	3	322
DOW 30	2000	4	353
Biotech	2001	1	37
Biotech	2001	2	45
Biotech	2001	3	11
DOW 30	2001	1	369
DOW 30	2001	2	372
DOW 30	2001	3	75

Slide 150

Knowledge of how the dimension values form contexts within the data will allow us to manipulate the contexts to form desired aggregate calculations.

For example, in a normalization report, we want to divide every number in the report by some total in a report. A report normalized by year, will have every number in the report divided by the total for each number's respective year, as in the formula: x / (Sum(x) for the entire year).

Context Operators

- Context operators allow us to manipulate the data contexts and override the default report contexts.
- BusinessObjects supplies four different operators
  - In: The In operator allows us to choose which dimensions in a report context will affect our calculations.
  - ForAll: The ForAll operator allows us to instruct BusinessObjects which dimensions in a report context will not affect our calculations.
  - ForEach: The ForEach operator allows us to define groups within the data context for our calculations.
  - Where: The Where operator allows us to override existing contexts and aggregate to specific dimensions values.
- There is an example of each operator in the following slides.

Slide 151

The context operators allow us to override and redefine the default contexts in a report that are due to report structure. Master of these operators will allow the creation of more complex and compact reports, especially since most of the operators allow for summaries within contexts.

The In Operator

- The In operator allows us to choose which dimension(s) in a report define the context for an aggregate function.
  - In this example, we chose the Portfolio Name dimension to define the context for our calculation.
    - Therefore, *Trans Year* and *Trans Quarter* no longer affect the calculation and it is only dependent on the values of the *Portfolio Name* object.
    - Therefore, in this example, the context in the Portfolio Total column is defined the same as the break footer.
  - The formula in the last column is a mixed context
    - = ([Num Transactions]) / ([Num Transactions] In ([Portfolio Name]))
    - The numerator calculates within the default context defined by the report
    - The denominator calculates within the context defined by the In operator, which is Portfolio Name.

Portfolio Name	Trans Year	Trans Quarter	Portfolio Total	Num Transactions	% Portfolio
DOW 30	2000	3	1,491	322	21.60 %
	2000	4	1,491	353	23.68 %
	2001	1	1,491	369	24.75 %
	2001	2	1,491	372	24.95 %
	2001	3	1,491	75	5.03 %
DOW 30			**Sum:**	**1.491**	**100.00 %**

Slide 152

Use the In Operator to Create a Percent Distribution

In Operator Example

1. Create a report from the SI Equity universe
2. From the Portfolio class, select *Portfolio Name*, *Trans Year*, *Trans Quarter* and *Num Transactions*
3. Click Run Query
4. Insert a break on *Portfolio Name*
5. Insert a sum on *Num Transactions*
6. Insert a column to the left of *Num Transactions*
7. Enter the following formula into the new column

   = [Num Transactions] In ([Portfolio Name])

   Notice that the value in each row is the same as the value in the footer. The In operator has created a new context for each row that is defined only by the value of the *Portfolio Name* dimension.
8. Insert a column to the right of *Num Transactions*
9. Enter the following formula

   = [Num Transactions] / [Num Transactions] In ([Portfolio Name])
10. Format the cell to Percent Style.

Note: We could do the percentages in the last column with the Percentage function. The formula will be: =Percentage ([Num Transactions])

In Operator Overrides Existing Report Contexts

- We use the In operator when we want to override existing contexts within a report
  - The default context for each row in this report is defined by Portfolio Name, Trans Year, and Trans Quarter
  - Each of the Report footer average calculations use a different context to define the sums before averaging
    - Avg Row: Default
    - Avg Portfolio: Portfolio Name
    - Avg Yearly: Trans Year
    - Avg Quarterly: Trans Quarter, Trans Year

Portfolio Name	Trans Quarter	Trans Year	Num Transactions
Alternative Energy	1	2001	49
	2	2001	36
	3	2000	42
	3	2001	11
	4	2000	36
Alternative Energy		Sum:	174
...	...		...
Technology	1	2001	133
	2	2001	131
	3	2000	126
	3	2001	53
	4	2000	148
Technology		Sum:	591
		Avg Row	89
		Avg Portfolio	447
		Avg Year	1,340
		Avg Quarter	536

Slide 153

Create a Summary with the In Operator

1. Create a report from the SI Equity universe
2. From the Portfolio class, select *Portfolio Name*, *Trans Year*, *Trans Quarter* and *Num Transactions*
3. Click Run
4. Insert a break on *Portfolio Name*
5. Insert a sum on *Num Transactions*
6. Insert three rows below the report footer row
7. Insert the following formulas into the rows

   = Average ([Num Transactions])

   = Average ([Num Transactions] In ([Portfolio Name]))

   = Average ([Num Transactions] In ([Trans Year]))

   = Average ([Num Transactions] In ([Trans Quarter]; [Trans Year]))

   For the Avg Row, Avg Portfolio, Avg Yearly, and Avg Quarterly, respectively.

In Operator Context Keywords

- Any number of dimensions can define a context for the In operator. However, if there is more than one, they must be separated by semicolons, and enclosed within parentheses. This is also true for the ForAll and ForEach operators.
  - In ([Trans Year]; [Portfolio Name])
- The In operator allows us to use keywords to define a context.
  - In Report: The entire report is the context
  - In Block (Section): The entire section of a master-detail is the context
  - In Body: The current default context for the formula. Rarely needed, because it is redundant to the default report logic.
  - In Break: The current break

Trans Year	Trans Quarter	Num Transactions	Num Transactions In Body	Num Transactions In Break	Num Transactions In Section	Num Transactions In Block	Num Transactions In Report
2000	3	42	42	78	174	174	2,680
	4	36	36	78	174	174	2,680
2000	Sum:	78					

Slide 154

Context keywords allow us to place general contexts on the In operator. This helps ensure that, if a report is reformatted, then the formulas will have a higher probability of still being accurate.

Section and Block are very similar, but Section ignores any filters placed on the section. Section is very useful in percentage reports.

In Block Example

- The In Block suffix allows us to override the default context of a table, crosstab, or chart, and defines the context as the entire section of a Section report.
  - This keyword only works in sections of Section reports
  - The formula in the % Trans column is
    - = [Num Transactions] / [Num Transactions] In Block
  - As opposed to
    - = [Num Transactions] / [Num Transactions] In ([Portfolio Name])
    - = [Num Transactions] / [Num Transactions] In ([Portfolio Industry])

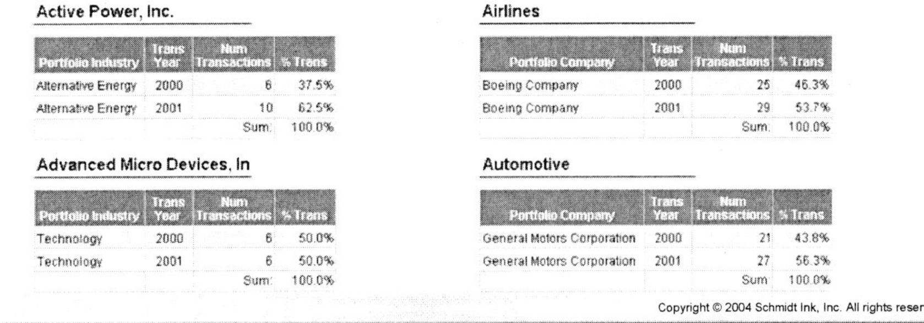

Active Power, Inc.

Portfolio Industry	Trans Year	Num Transactions	% Trans
Alternative Energy	2000	6	37.5%
Alternative Energy	2001	10	62.5%
		Sum:	100.0%

Airlines

Portfolio Company	Trans Year	Num Transactions	% Trans
Boeing Company	2000	25	46.3%
Boeing Company	2001	29	53.7%
		Sum:	100.0%

Advanced Micro Devices, In

Portfolio Industry	Trans Year	Num Transactions	% Trans
Technology	2000	6	50.0%
Technology	2001	6	50.0%
		Sum:	100.0%

Automotive

Portfolio Company	Trans Year	Num Transactions	% Trans
General Motors Corporation	2000	21	43.8%
General Motors Corporation	2001	27	56.3%
		Sum	100.0%

Slide 155

Calculation Using In Block

1. Create a report with *Portfolio Company*, *Portfolio Industry*, *Trans Year*, and *Num Transactions*
2. Right-click on any Portfolio Company and select *Set as Section*
3. Insert a column to the right of Num Transactions and insert the following formula

   = [Num Transactions] / [Num Transactions] In Block
4. Format the cell as a percent
5. Sum the new percent column (the sum should always be 100%)

The percentage formula in this example can be replaced with the Percentage function, as follows:

= Percentage ([Num Transactions])

The column is now relatively safe from online analysis. Which means that someone can now drag the Portfolio Industry object and drop it on the Portfolio Company master cell. This will cause the Portfolio Industry to now become the master. The % Trans column will correctly conform to the new context.

If the % Trans column was hard coded to use the Portfolio Company, then when Portfolio Industry became the master, the % Trans column would of become confusing and ambiguous.

Report Normalization with the In Operator

- The In operator enables us to view percent distribution across dimensions of a report
  - The middle table shows the percent distribution across the years in a report.
  - The table on the right shows the percent distribution across the entire report.

	2000	2001	Sum:
Alternative Energy	78	96	174
Biotech	91	93	184
DOW 30	675	816	1,491
Finance	45	75	120
Media	62	58	120
Technology	274	317	591
Sum:	1,225	1,455	2,680

	2000	2001
Alternative Energy	6 %	7 %
Biotech	7 %	6 %
DOW 30	55 %	56 %
Finance	4 %	5 %
Media	5 %	4 %
Technology	22 %	22 %
Sum:	100 %	100 %

Normal to Trans Year

	2000	2001	Sum:
Alternative Energy	3 %	4 %	6 %
Biotech	3 %	3 %	7 %
DOW 30	25 %	30 %	56 %
Finance	2 %	3 %	4 %
Media	2 %	2 %	4 %
Technology	10 %	12 %	22 %
Sum:	46 %	54 %	100 %

Normal to Report

Slide 156

Normalize Measure in a Report

1. Create a report from the SI Equity universe
2. From the Portfolio class, select *Portfolio Name*, *Trans Year*, and *Num Transactions*
3. Click Run Query
4. Format the table into an xtab with the Trans Year across the top
5. Create a copy of the crosstab.
6. Insert a sum on the original crosstab and view the values.
7. Select the copy and then click on the Properties tab. Select the Show Table Footers option in the Display section.
8. Make another copy of the crosstab.
9. Modify the measure formula to

    = [Num Transactions] / [Num Transactions] In ([Trans Year])

    This will normalize Num Transactions to Trans Year.
10. Copy the measure formula to the vertical footer.

    The total of each column should be 1 or 100%
11. Format the measure cells to percent
12. Modify the measure formula in the other crosstab to

    = [Num Transactions] / [Num Transactions] In Report

    This will normalize Num Transactions to the report
13. Copy the measure formula to all footer cells

    The column totals should be the percentage of transactions in each year

    The row totals are the percentage of transactions in each portfolio

    The report total should be 1 or 100%

Report Normalization with the Percentage Function

- The percentage function will allow us to create the same reports without the use of the context operator.
  - We learn the context operator, because it may be more flexible under certain circumstances.

	2000	2001	Sum:
Alternative Energy	78	96	174
Biotech	91	93	184
DOW 30	675	816	1.491
Finance	45	75	120
Media	62	58	120
Technology	274	317	591
Sum:	1.225	1,455	2.680

	2000	2001
Alternative Energy	6 %	7 %
Biotech	7 %	6 %
DOW 30	55 %	56 %
Finance	4 %	5 %
Media	5 %	4 %
Technology	22 %	22 %
Sum:	100 %	100 %

Normal to Trans Year

= Percentage ([Num Transactions]; Col)

	2000	2001	Sum:
Alternative Energy	3 %	4 %	6 %
Biotech	3 %	3 %	7 %
DOW 30	25 %	30 %	56 %
Finance	2 %	3 %	4 %
Media	2 %	2 %	4 %
Technology	10 %	12 %	22 %
Sum:	46 %	54 %	100 %

Normal to Report

= Percentage ([Num Transactions])

Slide 157

Normalize Measure in a Report with Percentage Function

1. Create a report from the SI Equity universe
2. From the Portfolio class, select *Portfolio Name*, *Trans Year*, and *Num Transactions*
3. Format the table into a crosstab with the Trans Year across the top
4. Copy the table and modify the measure formula to

   = Percentage ([Num Transactions]; Col)

   This will normalize Num Transactions to Trans Year.
5. Click on any percent and click the Insert Sum button.
   The Year totals will be 1
6. Format the measure cells to percent
7. The row totals and report total really do not make too much sense and they are usually deleted.
8. Copy the original crosstab and modify the measure formula to

   = Percentage ([Num Transactions])

   This will normalize Num Transactions to the report
9. Format the measure cells to percent
10. Sum the percentages, by clicking the *Insert Sum* button.

The ForAll Operator

- The ForAll operator allows us to choose which dimension(s) in a report will not be part of the definition of a context for an aggregate function.
  - In this example, we choose to ignore *Trans Year* and *Trans Quarter,* thus the remaining dimension, *Portfolio Name,* defines the context for the calculation.
    - Therefore, *Trans Year* and *Trans Quarter* no longer affect the calculation and it is only dependent on the values of the *Portfolio Name* object.
  - The formula in the last column is a mixed context
    - =[Num Transactions] / [Num Transactions] ForAll ([Trans Year] ; [Trans Quarter])
    - The numerator calculates within the default context defined by the report
    - The denominator calculates within the context defined by the ForAll operator

Portfolio Name	Trans Year	Trans Quarter	Portfolio Total	Num Transactions	% Portfolio
DOW 30	2000	3	1,491	322	21.60 %
	2000	4	1,491	353	23.68 %
	2001	1	1,491	369	24.75 %
	2001	2	1,491	372	24.95 %
	2001	3	1,491	75	5.03 %
DOW 30			Sum:	1,491	100.00 %

Slide 158

Create a Summary with the ForAll Operator

1. Create a report from the SI Equity universe
2. From the Portfolio class, select *Portfolio Name*, *Trans Year*, *Trans Quarter* and *Num Transactions*
3. Insert a break on *Portfolio Name*
4. Insert a sum on *Num Transactions*
5. Insert a column to the left of *Num Transactions*
6. Enter the following formula into the new column

   =[Num Transactions] ForAll ([Trans Year] ; [Trans Quarter])

   Notice that the value in each row is the same as the value in the footer. The ForEach operator has created a new context for each row that is defined only by the value of the *Portfolio Name* dimension, by ignoring the Trans Year and Trans Quarter values.
7. Insert a column to the right of *Num Transactions*
8. Enter the following formula

   = [Num Transactions] / [Num Transactions] ForAll ([Trans Year]; [Trans Quarter])
9. Format the cell to Percent Style.

The ForEach Operator

- The ForEach operator allows us to operate on the data context within a document
  - It allows us to define the grouping levels within an aggregate calculation
  - In this example, The ForEach operator allows us to determine the Average Quarterly Transactions by defining Trans Year – Trans Quarter Groupings within the Portfolio Name report context. In other words, the average of each year – quarter combination within each portfolio
    - = Average ([Num Transactions]
      ForEach ([Trans Year]; [Trans Quarter]))

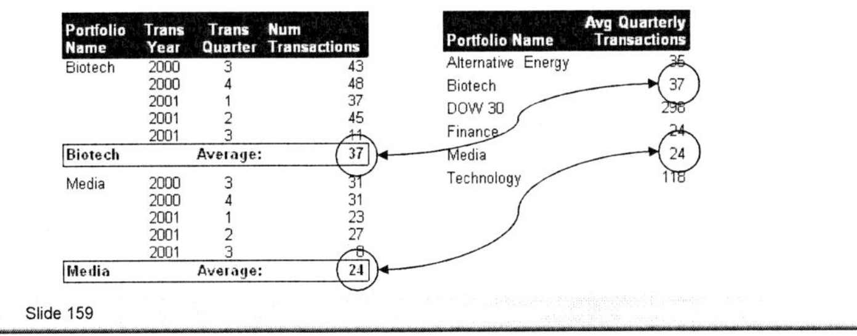

Portfolio Name	Trans Year	Trans Quarter	Num Transactions
Biotech	2000	3	43
	2000	4	48
	2001	1	37
	2001	2	45
	2001	3	11
Biotech		**Average:**	37
Media	2000	3	31
	2000	4	31
	2001	1	23
	2001	2	27
	2001	3	8
Media		**Average:**	24

Portfolio Name	Avg Quarterly Transactions
Alternative Energy	35
Biotech	37
DOW 30	298
Finance	24
Media	24
Technology	118

Portfolio Name	Avg Yearly Transactions
Alternative Energy	87
Biotech	92
DOW 30	746
Finance	60
Media	60
Technology	296

Slide 159

Create a Summary Report with the ForEach Operator

1. Create a report from the SI Equity universe
2. From the Portfolio class, select *Portfolio Name*, *Trans Year*, *Trans Quarter* and *Num Transactions*
3. Delete the *Trans Year* and *Trans Quarter* columns
4. Alter the formula in the *Num Transactions* column to

   = Average ([Num Transactions] ForEach ([Trans Year]; [Trans Quarter]))

5. Change the *Num Transactions* column header to

   Avg Quarterly Transactions

6. Insert a column to the right of the *Avg Quarterly Transactions* column
7. Enter the following formula

   = Average ([Num Transactions] ForEach ([Trans Year])

8. Enter the header

   Avg Yearly Transactions

Do you think that the following formula,
= Average ([Num Transactions] ForEach ([Trans Quarter])), would give desired results? Probably not. This would not give the average of the prior 5 quarters. It would yield the average of the total for each of the four quarter numbers.

ForEach Operator Defines Report Contexts

- The ForEach operator will add additional dimensions to the context definition
  - In the top report, the default context is Portfolio Name, Trans Year, and Trans Quarter
  - The average in each footer averages the Number of Transactions in each row of the footer, which is defined by the default context
  - Since the bottom table's default context is defined only by Portfolio Name, the average for each portfolio is simply its sum
    - Average (*Sum*(Num Transactions))
  - The ForEach operator allows us to add dimensions to the existing default context.
    - ForEach ([*Portfolio Name];* [Trans Year]; [Trans Quarter])

Portfolio Name	Trans Year	Trans Quarter	Num Transactions
Biotech	2000	3	43
	2000	4	48
	2001	1	37
	2001	2	45
	2001	3	11
Biotech			**37**
DOW 30	2000	3	322
	2000	4	353
	2001	1	369
	2001	2	372
	2001	3	75
DOW 30			**298**
Finance	2000	3	24
	2000	4	21
	2001	1	34
	2001	2	32
	2001	3	5
Finance			**24**
		Average:	**120**

Portfolio Name		Average Num Transactions
Biotech		37
DOW 30		298
Finance		24

Slide 160

To simply formulas, we usually try to leave out unnecessary elements, such as default calculation functions and default context dimensions.

The ForEach Operator Defines Granularity

- Year - Quarter resolution
  - = Average ([Num Transactions] ForEach ([Trans Year]; [Trans Quarter]))

Portfolio Name	Trans Year	Trans Quarter	Num Transactions
Biotech	2000	3	43
Biotech	2000	4	48
Biotech	2001	1	37
Biotech	2001	2	45
Biotech	2001	3	11

Portfolio Name	Num Transactions
Biotech	37
DOW 30	298
Finance	24

(43+48+37+45+11)/5 = 37

- Year Resolution
  - = Average ([Num Transactions] ForEach ([Trans Year]))

Portfolio Name	Trans Year	Trans Quarter	Num Transactions
Biotech	2000	3	43
Biotech	2000	4	48
Biotech	2001	1	37
Biotech	2001	2	45
Biotech	2001	3	11

Portfolio Name	Num Transactions
Biotech	92
DOW 30	746
Finance	60

(91+93)/2 = 92

Slide 161

Do you think that the following formula, *= Average ([Num Transactions] ForEach ([Trans Quarter]))*, would give desired results? Probably not. This would not give the average of the prior 5 quarters. It would yield the average of the total for each of the four quarter numbers.

Combining Context Operators

- Combining Context operators allows more complex formulas
  - For example, to calculate the variance from the portfolio mean, we could use the following formula
    - = [Num Transactions] -
      Average ([Num Transactions]
      ForEach ([Trans Year]; [Trans Quarter])) In ([Portfolio Name])
    - The ForEach operator instructs the Average function to average the totals for each Trans Quarter – Trans Year combination.
    - The In operator instructs the average function to average the combinations within each portfolio.
  - To calculate the variance from the report mean, we could use the following formula
    - = [Num Transactions] -
      Average ([Num Transactions]
      ForEach ([Trans Year]; [Trans Quarter])) In Report

Portfolio Name	Trans Year	Trans Quarter	Num Transactions	Variance from the Portfolio Mean	Variance from the Report Mean
Alternative Energy	2000	3	42	7.2	-47.33
Alternative Energy	2000	4	36	1.2	-53.33
Alternative Energy	2001	1	49	14.2	-40.33
Alternative Energy	2001	2	36	1.2	-53.33
Alternative Energy	2001	3	11	-23.8	-78.33

Portfolio Mean: 35
Report Mean: 89

Slide 162

People often talk about input and output contexts, which is probably confusing to most. It is probably better to think what the context operators are doing in the formula. For example, in this case, we want to calculate the average quarterly number of transactions, so we use Average ([Num Transactions] ForEach ([Trans Year]; [Trans Quarter])). The next question is: do we want this average for every portfolio or for the entire report. Once this decision is made, then we can use In <Portfolio Name> or In Report, respectively.

Difference from Average

1. Create a report with *Portfolio Name*, *Trans Year*, *Trans Quarter* and *Num Transactions*
2. Insert two columns to the right of Num Transactions.
3. In the first new column insert the formula
   = [Num Transactions] - Average ([Num Transactions] ForEach ([Trans Year]; [Trans Quarter])) In ([Portfolio Name])
4. In the second new column enter the formula
   = [Num Transactions] - Average ([Num Transactions] ForEach ([Trans Year]; [Trans Quarter])) In Report

In this example, the In operator allows us to override the existing natural contexts in the report. A natural context is defined by the dimensions in the report. In this case – Portfolio Name, Trans Year, and Trans Quarter. In In operator allows us to calculate for the portfolio context and the report context. The For Each operator allows us to average for the Year – quarter combinations within these two contexts.

The Where Operator

- The Where operator allows us to isolate measures to selected dimension values within a report context.
  - For example, the following formula will override any context in a report created by Portfolio Name. Thus causing Num Transactions to only aggregate for the portfolio DOW 30.
    = [Num Transactions]
    Where ([Portfolio Name]="DOW 30")

Portfolio Name	Total Transactions	DOW30 Transactions
Alternative Energy	174	1,491
Biotech	184	1,491
DOW 30	1,491	1,491
Finance	120	1,491
Media	120	1,491
Technology	591	1,491

  - The Where clause will not override other report contexts defined by other dimensions.
    - In the lower table, the Trans Year dimension has divided the DOW 30 number of transactions into 2000 and 2001 totals.

Portfolio Name	Trans Year	Total Transactions	DOW30 Transactions	% of DOW30 Transactions
Alternative Energy	2000	78	675	12%
Alternative Energy	2001	96	816	12%
Biotech	2000	91	675	13%
Biotech	2001	93	816	11%
DOW 30	2000	675	675	100%
DOW 30	2001	816	816	100%
Finance	2000	45	675	7%
Finance	2001	75	816	9%
Media	2000	62	675	9%
Media	2001	58	816	7%
Technology	2000	274	675	41%
Technology	2001	317	816	39%

Slide 163

Comparison Report Using the Where Operator

1. Create a report from the SI Equity universe
2. From the Portfolio class, select *Portfolio Name*, *Trans Year*, and *Num Transactions*
3. Insert two columns to the right of Num Transactions
4. Rename the Num Transactions header to Total Transactions
5. Insert the following formula into the next column
   = [Num Transactions] Where ([Portfolio Name] = "DOW 30")
6. Insert the following into the header
   DOW30 Transactions
7. Insert the following formula into the remaining column
   =[Num Transactions] / ([Num Transactions] Where ([Portfolio Name] = "DOW 30"))
   If the above formula does not calculate properly, use the following fix:
   = ToNumber([Num Transactions]) / ToNumber([Num Transactions]) Where ([Portfolio Name] = "DOW 30")
   (This formula was proposed to me by Stan Montwedi from Datamode Systems)
8. Insert the following into the header
   % of DOW30 Transactions

There is also a good example of the Where operator on slide: Numeric Functions – ToNumber (MTD Report).

Rest of World Example

- The Where operator allows us to isolate selected dimension values. However, when used in a formula, it can also allow us to isolate the rest of the world.
  - = [Num Transactions] Where ([Portfolio Name] = "DOW 30") allows us to isolate the number of DOW 30 transactions
  - = [Num Transactions] – ([Num Transactions] Where ([Portfolio Name] = "DOW 30")) allows us to subtract the DOW 30 transactions from the total transactions, thus isolating the rest of the world values.

Trans Year	Trans Quarter	Total Transactions	DOW30 Transactions	Rest of Transactions	% DOW30 Transactions
2000	3	588	322	266	55 %
2000	4	637	353	284	55 %
2001	1	645	369	276	57 %
2001	2	643	372	271	58 %
2001	3	167	75	92	45 %

Slide 164

Rest of World Report Using the Where Operator

1. Create a report from the SI Equity universe
2. From the Portfolio class, select *Portfolio Name*, *Trans Year*, *Trans Quarter* and *Num Transactions*
3. Insert three columns to the right of Num Transactions
4. Delete the Portfolio Name column
5. Change the Num Transactions header to Total Transactions
6. Enter the following formula into the next column
   = [Num Transactions] Where ([Portfolio Name] = "DOW 30")
7. Change the header to
   DOW 30 Transactions
8. Enter the following formula into the next column
   = [Num Transactions] - ([Num Transactions] Where ([Portfolio Name] = "DOW 30"))
   If the above formula does not calculate properly, use the following formula
   = ToNumber([Num Transactions]) - ToNumber([Num Transactions]) Where ([Portfolio Name] = "DOW 30")
   (This formula was proposed to me by Stan Montwedi from Datamode Systems)
9. Change the header to
   Rest of Transactions
10. Enter the following formula into the remaining column
    = ([Num Transactions] Where ([Portfolio Name]="DOW 30")) / [Num Transactions]
    If the above formula does not calculate properly, use the following formula
    = ToNumber([Num Transactions]) Where ([Portfolio Name]="DOW 30") / ToNumber([Num Transactions])
    (This formula was proposed to me by Stan Montwedi from Datamode Systems)
11. Change the header to
    % DOW 30 Transactions

At the time of this writing, some of the formulas in this example were not calculating properly. The last two columns were returning strange results. I am including this example hoping that this will be fixed in future releases.

Where Operator Restrictions

- The Where operator only allows the following operators
  - Between, Inlist, =
    =[Num Transactions] Where ([Trans Year]=2000)
    - It does not allow: < >, <=, >=, <, >, …
  - logic AND, OR, NOT
    =<Num Transactions>
    Where ([Trans Year]=2000 AND [Trans Quarter] = 1)
- We can use more complicated logic, if we encapsulate it within a variable that uses the If function.
  - For example, a variable named *First Half* may have the following formula: = If ([Trans Quarter] InList (1, 2); 1; 0)
  - Then the Where clause would be
    = [Num Transactions] Where ([First Half] = 1)

Slide 165

Many people get frustrated, because of the logic limitations on the Where operator. However, as shown above these limitations are easily overcome with the use of the If function. Most report developers that I know, always use the If function, even for simple logic, such as [Trans Year] = 2000. They do this because it encapsulates all their logic into easy to find and modifiable variable packages with names, such as IsYear2000, IsProduction, IsValidClient, and so forth.

Prompted Rest of World Example

- In this example, we are going to modify the previous Rest of World example to use prompted input to select the portfolio of interest.
- Since the Where operator will not work with functions, we will use If-Then logic encapsulated in a variable. The variable will have the following
  - Name: IsPortfolioName
  - Formula:
    = If (UserResponse("Enter Portfolio") = [Portfolio Name]; 1; 0)

Trans Year	Trans Quarter	Total Transactions	Biotech Transactions	Rest of Transactions	% Biotech Transactions
2000	3	588	43	545	7%
2000	4	637	48	589	8%
2001	1	645	37	608	6%
2001	2	643	45	598	7%
2001	3	167	11	156	7%

Slide 166

Where Operator Using Advanced Logic

1. Modify the query of the previous report to contain the following condition

   Portfolio Name Less than Prompt ('Enter Portfolio')

   Or

   Portfolio Name Greater than or equal to Prompt ('Enter Portfolio')
2. Click Run and select any portfolio
3. Create the following variable

   Name: IsPortfolioName

   Formula: = If (UserResponse("Enter Portfolio") = [Portfolio Name]; 1; 0)
4. Change the DOW 30 Transactions header to

   = UserResponse("Enter Portfolio") + " Transactions"
5. Enter the following Formula into the column

   = [Num Transactions] Where ([IsPortfolioName] = 1)
6. In the Rest of Transactions column, enter the following formula

   = ToNumber([Num Transactions]) - ToNumber([Num Transactions]) Where ([IsPortfolioName] = 1)
7. Change the column header in the last column to

   = "% " + UserResponse("Enter Portfolio") + " Transactions"
8. Enter the following formula into the column

   = ToNumber([Num Transactions]) Where ([IsPortfolioName]=1) / ToNumber([Num Transactions])

Data and Report Contexts Summary

- In this chapter
  - We learned that dimensions in a report create contexts. We also learned that we can manipulate these contexts with context operators and keywords.
  - We learned how to create several very powerful reports that take advantage of our knowledge of contexts and context operators.

Slide 167

This chapter introduced us to a new way of organizing data – with contexts. These contexts allow us to create powerful and informative reports.

Slide 168

Creating Documents with Web Intelligence XI

Sorts, Filters, Ranks, and Alerters (Java Report Panel)

Slide 169

Introduction

- Now that we can create report structures, we can enhance our reports by sorting, filtering, and ranking the results. We can also apply alerters that allow us to emphasize information based on the values in a report.
- In this chapter we will learn how
  - To place, remove, modify, and locate sorts in a report.
  - To filter data so that only the information of interest is visible.
  - To filter or rank values on their relative strengths in a document.
  - To place alerters that emphasizes the information in a report.

Slide 170

Sorts

- Sorts allow us to arrange information in descending or ascending order. We can even, create custom sorts.
- We can sort the
  - Rows of a table by any combination of columns.
  - Rows and/or columns of a crosstab. We can also sort the body.
  - Chart axis values and the magnitude of the graphic elements.

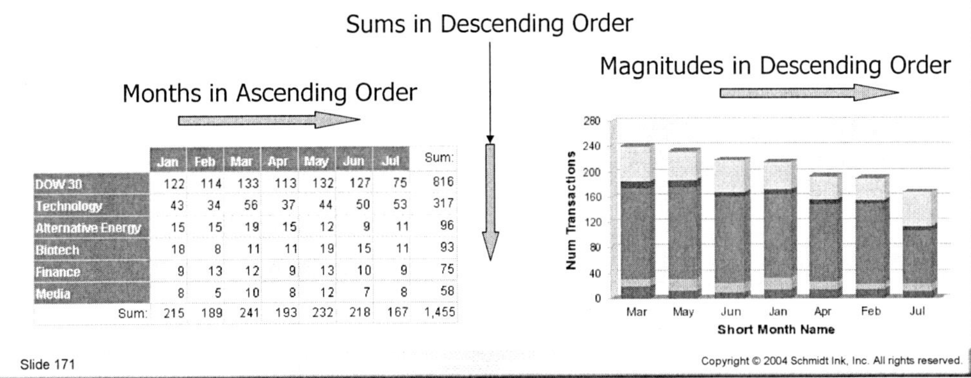

	Jan	Feb	Mar	Apr	May	Jun	Jul	Sum:
DOW 30	122	114	133	113	132	127	75	816
Technology	43	34	56	37	44	50	53	317
Alternative Energy	15	15	19	15	12	9	11	96
Biotech	18	8	11	11	19	15	11	93
Finance	9	13	12	9	13	10	9	75
Media	8	5	10	8	12	7	8	58
Sum:	215	189	241	193	232	218	167	1,455

Slide 171

We can sort almost any aspect of a report. BusinessObjects is very flexible in this way, although it may not always be obvious how to accomplish the sort. In this chapter, we will talk about the various ways to apply a sort.

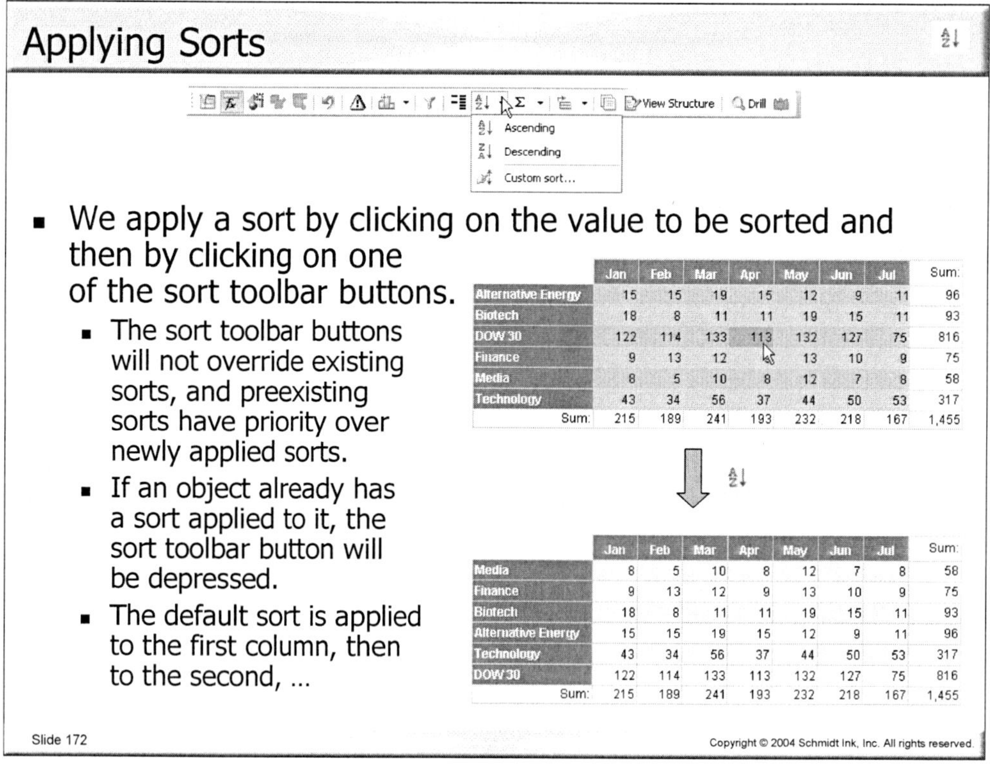

	Jan	Feb	Mar	Apr	May	Jun	Jul	Sum:
Alternative Energy	15	15	19	15	12	9	11	96
Biotech	18	8	11	11	19	15	11	93
DOW 30	122	114	133	113	132	127	75	816
Finance	9	13	12	[illegible]	13	10	9	75
Media	8	5	10	8	12	7	8	58
Technology	43	34	56	37	44	50	53	317
Sum:	215	189	241	193	232	218	167	1,455

	Jan	Feb	Mar	Apr	May	Jun	Jul	Sum:
Media	8	5	10	8	12	7	8	58
Finance	9	13	12	9	13	10	9	75
Biotech	18	8	11	11	19	15	11	93
Alternative Energy	15	15	19	15	12	9	11	96
Technology	43	34	56	37	44	50	53	317
DOW 30	122	114	133	113	132	127	75	816
Sum:	215	189	241	193	232	218	167	1,455

The Sort toolbar buttons allows us a quick way to sort an object. However, they do not override existing sorts on other objects in the same report. Therefore, the buttons will only allow sorts within existing sorts in a report.

Sorting a Table

1. Create a Report with Portfolio Name, Trans Month Name, and Num Transactions.
   To create the short names used in the example, create a variable with the formula
   = Left ([Trans Month Name]; 3)
2. Create a crosstab with Trans Month Name as the Column Header.
3. Click on any Num Transactions value in the body of the crosstab.
4. Click the Insert Sum toolbar button.
5. Click the small down arrow to the right of the Apply/Remove Sort toolbar button.
6. Click on the Descending Sort button.
7. Click on any Portfolio Name in the Row Header column.
8. Click the Ascending Sort option in the Reporting toolbar.
   (Nothing will happen, since a sort has already been applied to Num Transactions)

Viewing and Modifying Applied Sorts

- Many times, we are not sure what sorts are placed on a table. To see existing sorts and their sort priority, select the table and then click on the Properties tab in the Manager. Open the Sorts section and click the little button that appears when the cursor is placed to the left of the Sort Priority option.
  - To delete a sort, click on it in the list and press the [Delete] key.
  - To reorder the sort priority, highlight one of the sorts and move it up or down in the list.
  - Notice crosstabs have vertical and horizontal sorts.

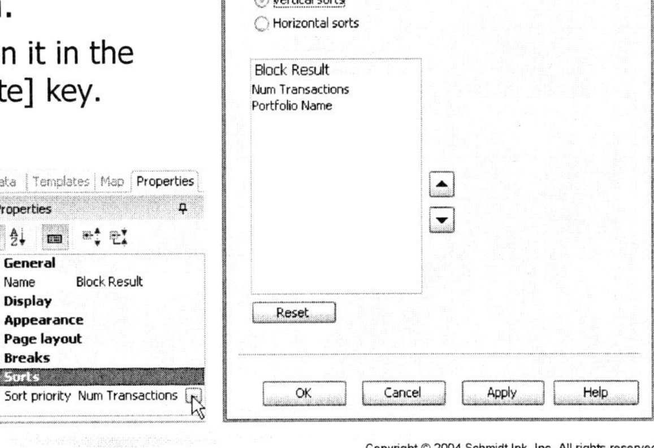

Slide 173

Since report structures can have many sorts applied to them, it is convenient to use the Sorts dialog to view all of the sorts placed on a structure. The sorts are in their order of priority. In this example, the Num Transactions sort first, then the Portfolio Name object.

Crosstabs and certain charts have both vertical and horizontal sorts. The vertical sorts sort the structure up and down – along the row headers. The horizontal sorts sort a structure right to left – along the column headers.

We can delete a sort in the list by highlighting it and then pressing the [Delete] key.

Sorts in the Sorts Dialog

1. Select the crosstab we created in the previous example.
2. Click on the Properties tab of the Manager.
3. Open the Sorts section.
4. Place the cursor in the cell to the right of the Sort Priority option and click the little button that appears in the cell.
5. Make sure that the Vertical Sorts option is selected.
6. Highlight Portfolio Name in the list.
7. Click the up button to move it above Num Transactions. (Portfolio Name will now sort before Num Transactions)
8. Since there is only one Total Num Transaction per Portfolio Name, we do not need the sort on Num Transactions. Select Num Transactions in the list and press the [Delete] key to remove it from the list.
9. Click Ok to apply the new sort priority.

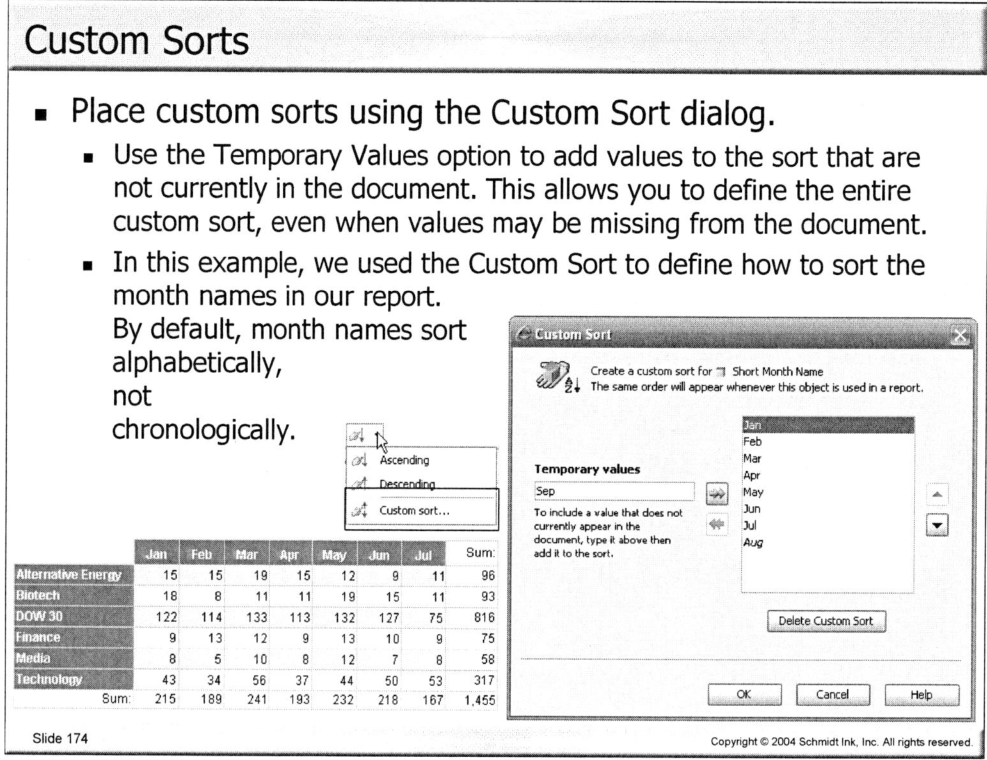

	Jan	Feb	Mar	Apr	May	Jun	Jul	Sum:
Alternative Energy	15	15	19	15	12	9	11	96
Biotech	18	8	11	11	19	15	11	93
DOW 30	122	114	133	113	132	127	75	816
Finance	9	13	12	9	13	10	9	75
Media	8	5	10	8	12	7	8	58
Technology	43	34	56	37	44	50	53	317
Sum:	215	189	241	193	232	218	167	1,455

Many times there is a need to define the sort order for the values of an object. Companies may like to see the departments listed in a certain order, they may want their products listed from most significant to least, or the names of months sorted not in alphabetical order, but in chronological order.

Notice that we can delete Custom Sorts in the Custom Sort dialog.

Applying a Custom Sort

1. Using the previous example, click on any of the Trans Month Names in the column header. If you used the Left ([Trans Month Name]; 3) formula, you will have define it as a variable.
2. Click the Custom Sort option in the Reporting toolbar.
3. Arrange the month values in the list to be in chronological order.
4. Add the missing months with the Temporary Values edit option.
5. Click OK.

Filters

- Filters allow us to make selected information in a report or report structure visible, while hiding other information in the report. This allows us to
  - Isolate information in large reports.
  - Perform calculations that compare visible information with hidden information.
  - Have a large query that returns a superset of data, and then display subsets of the information in different report structures.
- Filters can not be placed on aggregated data.
  - Data that has been summed, counted, averaged, and so forth.

Portfolio Name	Trans Year	Revenue/ Expense
Alternative Energ	2000	-89,221
Alternative Energ	2001	303,886
Biotech	2000	-369,888
Biotech	2001	449,314
DOW 30	2000	-4,385,413
DOW 30	2001	5,736,866
Finance	2000	-507,101
Finance	2001	574,563
Media	2000	-465,394
Media	2001	386,854
Technology	2000	-1,116,767
Technology	2001	1,059,497

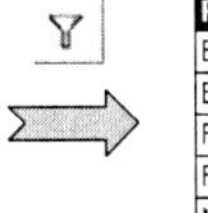

Portfolio Name	Trans Year	Revenue/ Expense
Biotech	2000	-369,888
Biotech	2001	449,314
Finance	2000	-507,101
Finance	2001	574,563
Media	2000	-465,394
Media	2001	386,854

Slide 175

Many companies use filters as often as sorts. They use them to hide information that is necessary for calculations in a report, but not necessary to display in a report. They also use them for many other reasons, therefore it is important to become very proficient at creating and placing filters.

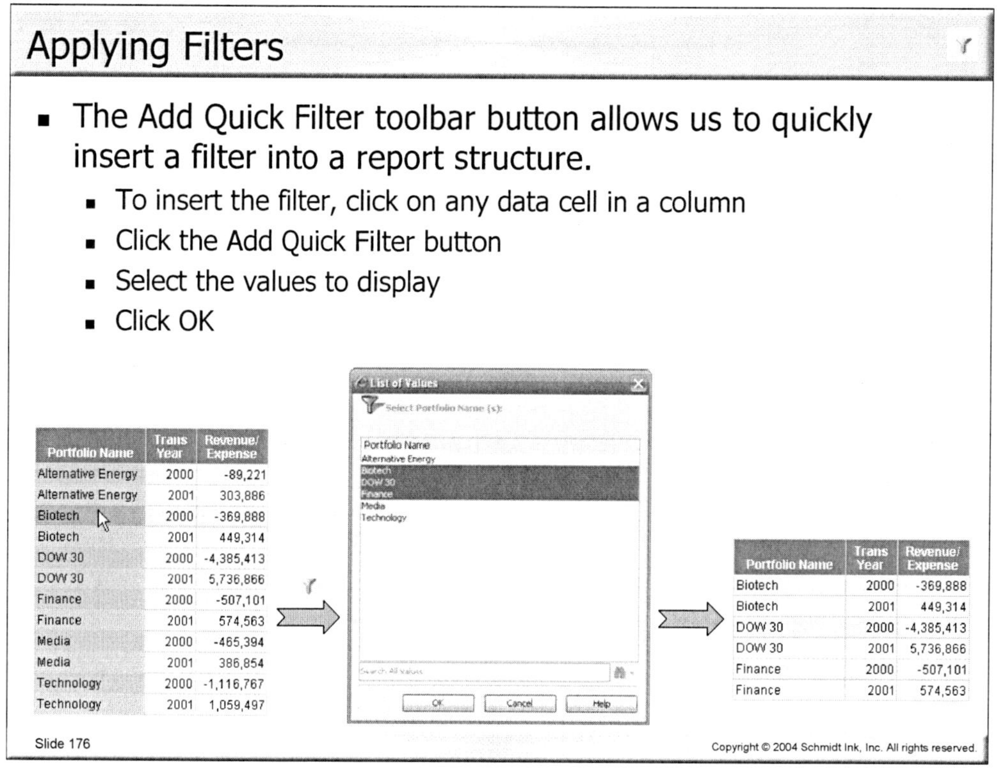

The List of Values for filters is different from the list of values for Query Filters. For Query Filters, it lists all of the available values in the database. For Report Filters, it lists all of the available values in the document.

Inserting a Simple Filter

1. Create a report with Portfolio Name, Trans Year, and Revenue/ Expense.
2. Click on any of the Portfolio Name values.
3. Click the Add Quick Filter toolbar button.
4. Select Biotech, DOW 30, and Finance..
5. Click OK

Note: The #EMPTY value in the list represents all unassigned values. This simply means that there may be data in the report where the row has no Portfolio Name.

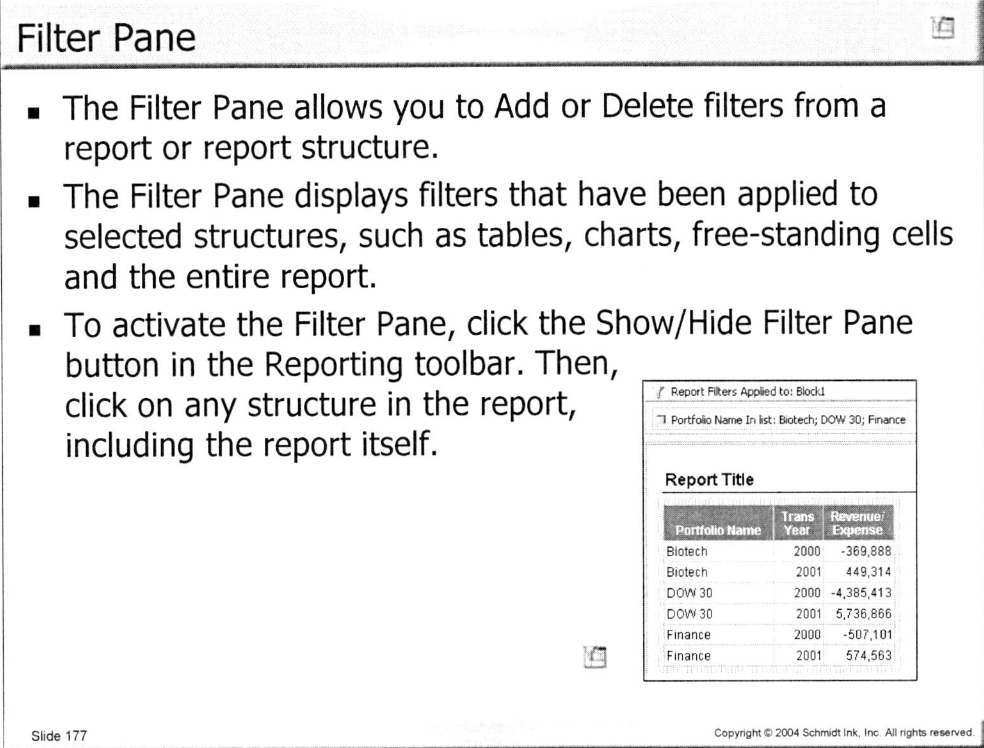

Filter Pane
The Filter Pane allows you to Add or Delete filters from a report or report structure.
The Filter Pane displays filters that have been applied to selected structures, such as tables, charts, free-standing cells and the entire report.
To activate the Filter Pane, click the Show/Hide Filter Pane button in the Reporting toolbar. Then, click on any structure in the report, including the report itself.
Report Filters Applied to: Block1
Portfolio Name In list: Biotech; DOW 30; Finance
Report Title
Portfolio Name
Trans Year
Revenue/ Expense
Biotech 2000 -369,888
Biotech 2001 449,314
DOW 30 2000 -4,385,413
DOW 30 2001 5,736,866
Finance 2000 -507,101
Finance 2001 574,563
Slide 177
Copyright © 2004 Schmidt Ink, Inc. All rights reserved.

The Filter Panel is very convenient, because it allows us to quickly view any filters placed on selected report structures.

View Report Filter on Selected Table

1. Click anywhere on the table from the previous exercise.
   (In this exercise, we placed a filter on Portfolio Name in the table)
2. Click the Show/Hide Filter Pane button in the Reporting toolbar.
   (Notice the filter *Portfolio Name In List Biotech;DOW 30;Finance*.

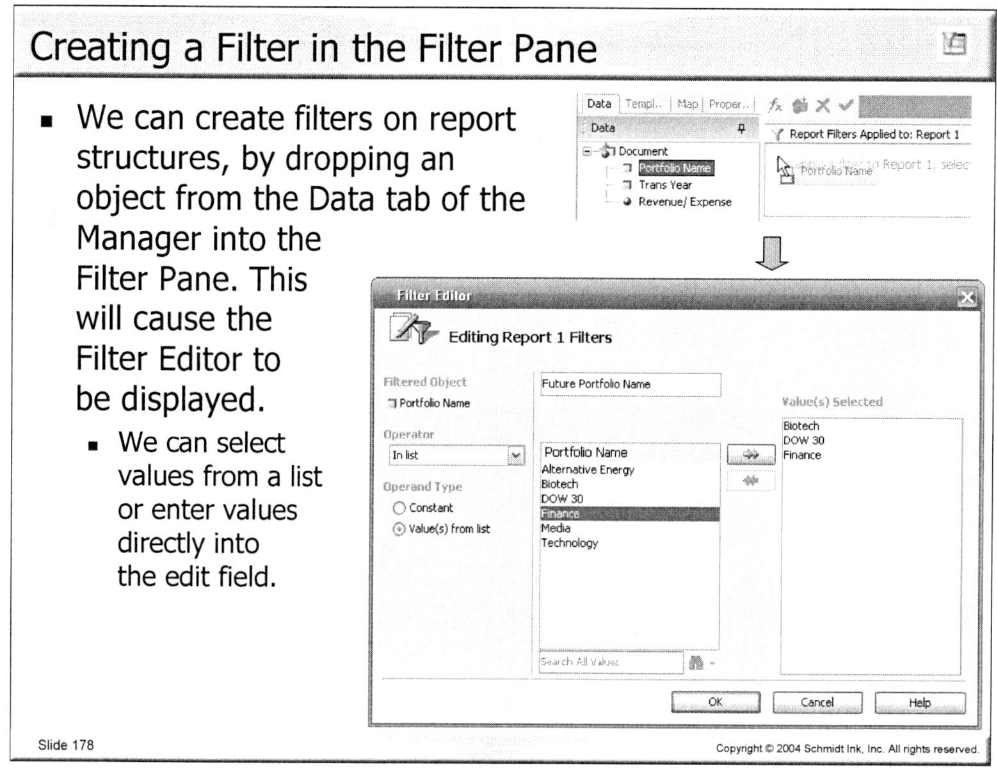

We can create Report Filters on selected structures by showing the Filter Pane and then dropping objects into the pane. When an object is dropped into the Filter Pane, the Filter Editor is displayed. In this editor, we can select an operator and select values for the operand. The operand values can be selected from the list of values and/or entered into the edit field. In either case, all values must appear in the Value(s) Selected list. To get a typed value from the edit field to the list, just press the [Enter] key after typing the value.

Create a Filter Using the Filter Pane

1. Create a report with Portfolio Name, Trans Year, and Num Transactions.
2. Click the Show/Hide Filter Pane button to display the filter pane.
3. Drag the Portfolio Name object from the Data tab and drop it into the Filter Pane.
4. Select the In List operator from the drop-down list.
5. Select the Value(s) from list option.
6. Double-click on Biotech, DOW 30, and Finance.
7. Type Future Portfolio Name into the Edit Field near the top of the dialog and press the [Enter] key to accept the value.
8. Click OK to apply the filter.

Using a Formula in the Filter Pane

- To use a formula in the Filter Pane, we simply create a variable with the filter logic and then apply the filter to the variable.
  - For example, to create a formula that hides all rows with null values, you create a variable similar to the following
    - Name: IsNumCallsNull
    - Formula: =If (IsNull ([Num Calls]);0;1)
  - Then in the Filter Pane, do the following

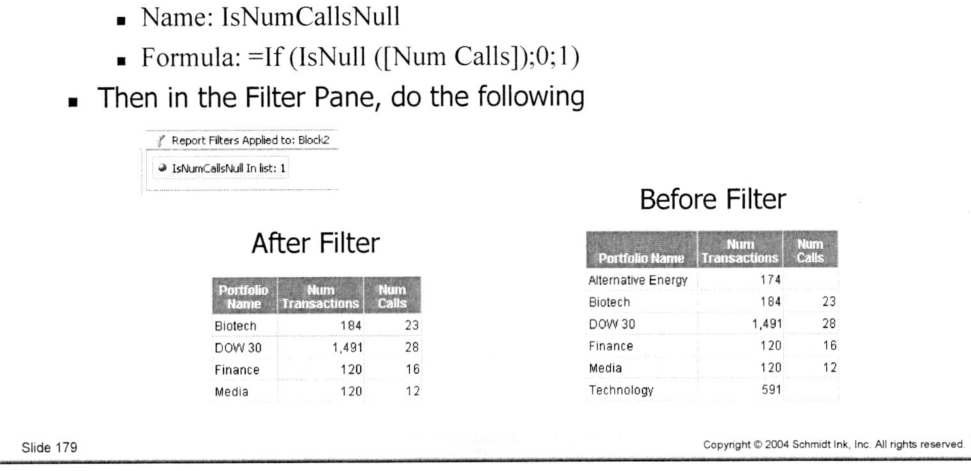

After Filter

Portfolio Name	Num Transactions	Num Calls
Biotech	184	23
DOW 30	1,491	28
Finance	120	16
Media	120	12

Before Filter

Portfolio Name	Num Transactions	Num Calls
Alternative Energy	174	
Biotech	184	23
DOW 30	1,491	28
Finance	120	16
Media	120	12
Technology	591	

Slide 179

Many times we want to apply a filter that is more complicated, then simply choosing values from a list. These filters need a formula to define their behavior. The If function allows us to create complicated filter logic, because it is either true or false – either it is to be filtered or it is not.

Create a Filter Formula

1) Create a report with Portfolio Name and Num Transactions.
2) Edit the Query and Add a query for Call Portfolio Name and Num Calls.
3) Merge the Portfolio Name objects.
4) Create a table with Portfolio Name, Num Transactions, and Num Calls. (Two of the Portfolios never made any calls. Therefore, the Num Calls for both portfolios is null.)
5) Insert a column to the right of Num Calls and enter the following formula = If (IsNull ([Num Calls]);0;1)
6) Click the Create Variable button in the Formula toolbar and name the new formula – IsNumCallsNull.
7) Type it as a measure and click OK.
8) Click the Show/Hide Filter Pane button to display the filter pane.
9) Drag the IsNumCallsNull object from the Data tab in the Manager and drop it into the Filter Pane.
10) Select the Equal to operator and type a 1 into the Type a value edit field.
11) Click the OK button.

Advanced note: If the IsNumCallsNull variable were typed as a dimension, the filter may not be necessary, because the table will only show rows where there is a value of 1. The reason for this is that the dimension must have merged values to make it acceptable to the table. To check this…

1) Delete the filter
2) Double-click on the IsNumCallsNull object in the Data tab of the Manager.
3) Change the type from a measure to a dimension in the Variable Editor.
4) Click OK.

Using the NoFilter Function

- The NoFilter function allows us to ignore filters that have been placed on a report structure or report.
- To use the function, we simply assign a formula as its argument.
- Many times, after a filter has been placed on a report structure, we want to know the percentage that the filtered values represent of the entire population. We can do this with a formula similar to
  - = NoFilter([Num Transactions]/ [Num Transactions] In Report)
  - Or, = NoFilter(Percentage ([Num Transactions]))

Portfolio Name	Num Transactions	% Total Transactions
Biotech	184	7%
DOW 30	1,491	56%

Slide 180

The NoFilter function is very valuable, because it allows us to ignore filters placed on a report. This allows us to access all of the data in a document to create formulas that can compare the visible values to the total of all values.

Use the NoFilter Function

1) Create a report with Portfolio Name and Num Transactions.
2) Place a filter that shows only the DOW 30 and Biotech portfolios.
3) Insert a column to the right of Num Transactions.
4) Enter the following formula:
   = NoFilter([Num Transactions]/ [Num Transactions] In Report)

Ranks

- Ranks are special filters that allow us to filter dimension values based on the ordinal position of a measure relative to the dimension.
  - For example, we can rank the top three portfolios based on the total revenue in a report, as shown below.
  - We can also rank the bottom performers.
  - We can rank by values, percentage of total number of values (percentiles), Cumulative Sum, and Cumulative Percentage.

Portfolio Name	Trans Year	Revenue/ Expense
Alternative Energy	2000	-89,221.3
Alternative Energy	2001	303,885.9
Biotech	2000	-369,888.4
Biotech	2001	449,314.1
DOW 30	2000	-4,385,412.5
DOW 30	2001	5,736,865.5
Finance	2000	-507,100.7
Finance	2001	574,563
Media	2000	-465,393.9
Media	2001	386,853.6
Technology	2000	-1,116,767.25
Technology	2001	1,059,497.45
	Sum:	**1,577,195.5**

Top three portfolios

Portfolio Name	Trans Year	Revenue/ Expense
DOW 30	2000	-4,385,412.5
DOW 30	2001	5,736,865.5
Alternative Energy	2000	-89,221.3
Alternative Energy	2001	303,885.9
Biotech	2000	-369,888.4
Biotech	2001	449,314.1
	Sum:	**1,645,543.3**

Slide 181

Ranks are special filters that filter based on the top or bottom values of a measure that is related to the ranked dimension.

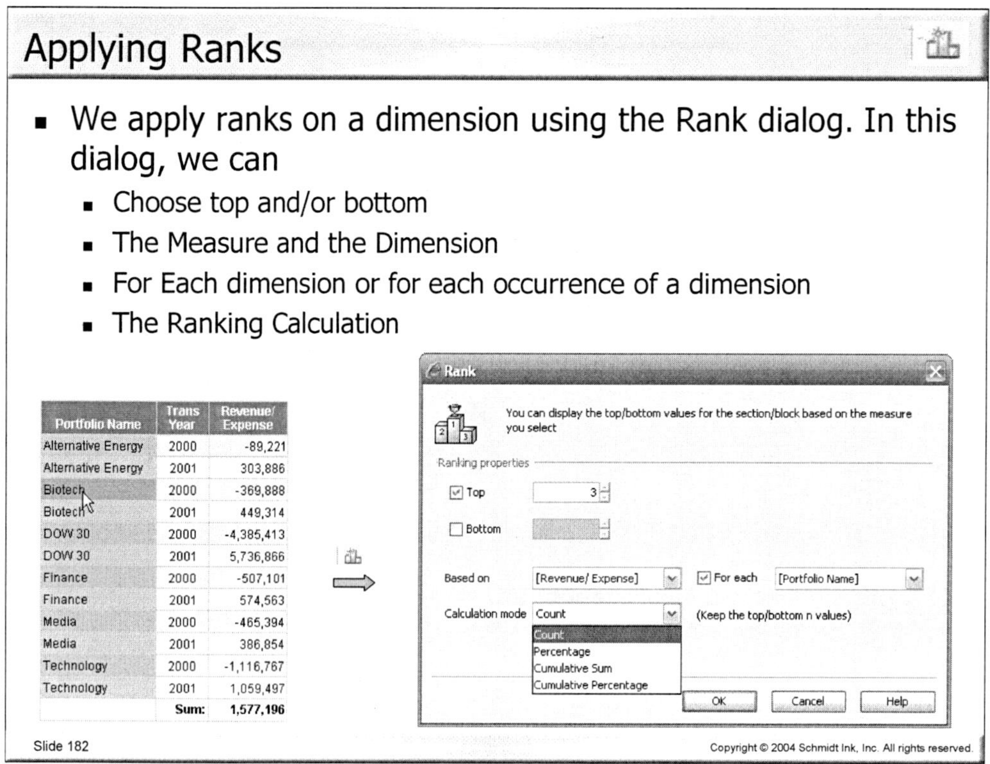

When we define a rank, we can choose the top, the bottom, or both the top and the bottom ranked dimensions. We rank dimensions with measures. For example, a dimension with the highest measure value could be called the top ranked dimension.

When we rank dimensions, we can rank them independent of the context in the report. This means that we will rank the dimensions on their measure's total value in the report, break or section. To calculate the rank in this manner, we choose the For Each option, which tells the rank operation to sum the measure for each dimension value before ranking.

We can also rank a dimension within its context – without summing all of the measure values to get a total before ranking. This allows us to basically rank the rows in a table, section, or break.

Applying a Rank to a Dimension

1. Create a report with Portfolio Name, Trans Year, and Revenue/Expense.
2. Click on any Portfolio Name in the table to select it.
3. Click the Apply/Remove Ranking toolbar button.
4. Select the Top option and enter a three into the edit field.
5. Keep the rest of the defaults and click OK.
   You should have six rows with DOW 30, Alternative Energy, and Biotech. These are the top three portfolios in total Revenue.
6. Click the down arrow to the right of the Apply/Remove Ranking button and select Edit Ranking.
7. Clear the For Each option and Click OK.
8. Now, we have three rows with DOW 30 – 2001, Technology- 2001, and Finance – 2001, These are the top three rows with the highest Revenue.

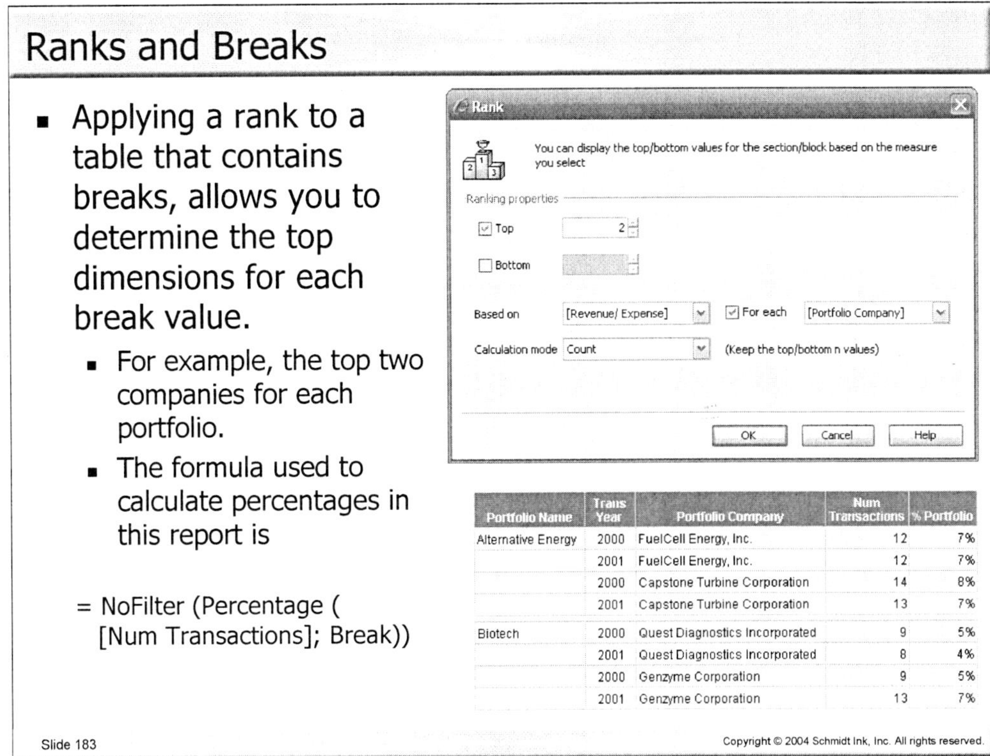

Portfolio Name	Trans Year	Portfolio Company	Num Transactions	% Portfolio
Alternative Energy	2000	FuelCell Energy, Inc.	12	7%
	2001	FuelCell Energy, Inc.	12	7%
	2000	Capstone Turbine Corporation	14	8%
	2001	Capstone Turbine Corporation	13	7%
Biotech	2000	Quest Diagnostics Incorporated	9	5%
	2001	Quest Diagnostics Incorporated	8	4%
	2000	Genzyme Corporation	9	5%
	2001	Genzyme Corporation	13	7%

Applying a Rank to a Table with Breaks

1. Create a report with Portfolio Name, Trans Year, Portfolio Company, Revenue/Expense, and Num Transactions.
2. Apply a break to Portfolio Name.
3. Click on any Portfolio Company and click the Apply/Remove Rank button.
4. Enter a 2 in the Top edit field.
5. Make sure it is based on Revenue/Expense For Each Portfolio Company.
6. Click OK.
7. Insert a column to the right of Num Transactions and insert the following formula
   = NoFilter (Percentage ([Num Transactions]; Break))
   In this formula, the NoFilter tells Business Objects to calculate the percentage using all of the companies, not just the ranked ones. The Break operator tells Business Objects to calculate the percentage with just the companies within each break.
8. Insert a column to the right of Revenue/ Expense and insert the following formula
   = NoFilter (Percentage ([Revenue/ Expense]; Break))
9. Copy all calculated fields to the break footer.

Alerters

- Alerters allow us to highlight or modify information in a report based on a logical formula.
  - We may highlight rows in a table where the revenue is less than zero, as shown below.
  - We can replace values with text or values from another object. We can even replace the contents of a cell with a formula.
  - We can format any attribute of a cell based on values in a report, including the font, shade, justification, and borders.

Portfolio Name	Revenue/ Expense
Alternative Energy	214,665
Biotech	79,426
DOW 30	1,351,453
Finance	67,462
Media	-78,540
Technology	-57,270

Slide 184

Alerters are used in many ways in different companies. They are flexible and allow people to get creative when applying them.

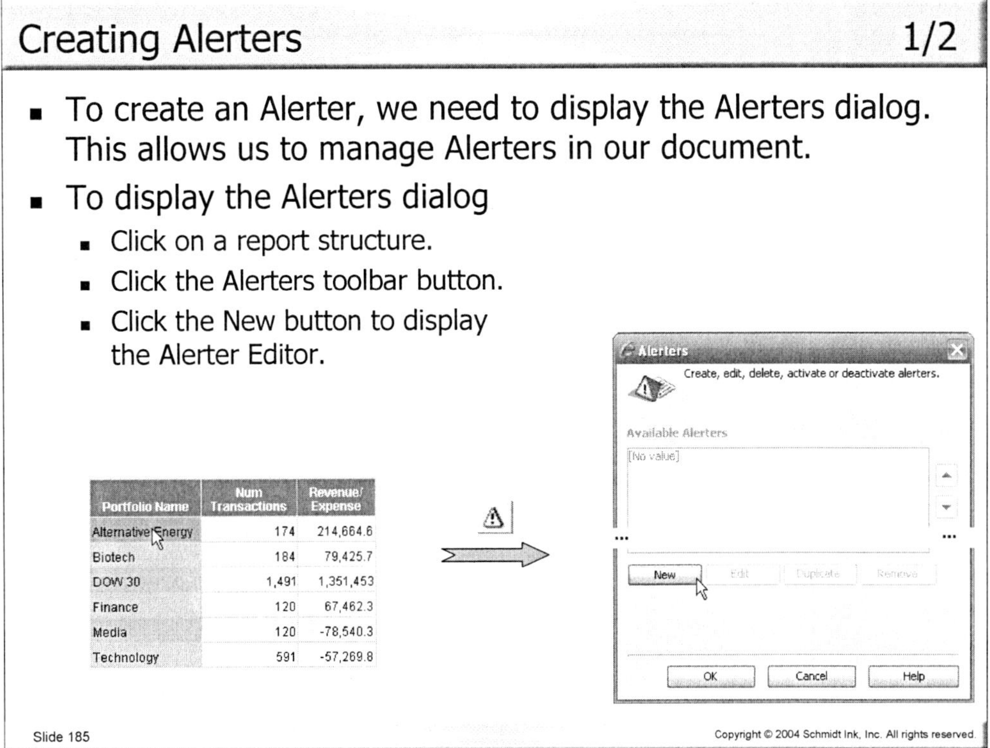

The Alerters dialog displays all of the alerters in a document. This allows you to edit, delete, or apply existing alerters to report structures in any report in a document.

Applying an Alerter

1. Create a report with Portfolio Name, Num Transactions, and Revenue/ Expense.
2. Click any value in the Portfolio Name column.
3. Click the Alerters toolbar button.
4. Click the *New..* Button to display the Edit Alerters dialog.
5. Continue on next page….

Creating Alerters 2/2

- The Add... button displays the Edit Alerters dialog.
  - With this dialog, we create new Alerters and modify existing ones.
- The definition tab allows us to name and describe an alerter.
  - The name and description of the alerter help us to determine the purpose of an alerter. In a report with many alerters it is very necessary to fill in these fields to help others to understand the alerters in the report.

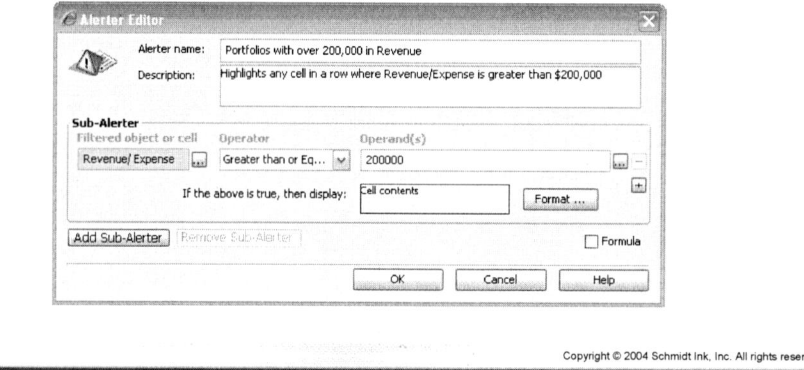

Slide 186

The Definition tab allows us to name and describe alerters. This is very important, if we want to remember how our alerters function - without having to examine the logic each time we revisit our alerters. It is also good to give a description to let others know the purpose of an alerter.

1. Enter *Portfolios with over 200,000 in Revenue* into the Alerter Name edit field.
2. Enter *Highlights any cell in a row where Revenue/Expense is greater than $200,000* into the Description field.
3. Enter the object for the Alerter comparison
    1. Click the small button the right of the Filtered Object or Cell field.
    2. Click on the Select Object or Variable entry on the pop-up menu.
    3. Select Revenue/ Expense from the Objects and Variables dialog.
    4. Click OK.
4. Select Greater than or Equal to from the Operator drop-list.
5. Click in the Operand edit field and type 200000 (with no commas).
6. Click the Format... button
    1. Set the font color to Default
    2. Set the Background color to light yellow
    3. Set the border color to blue.
    4. Click OK
7. Click OK
8. Continue on next page...

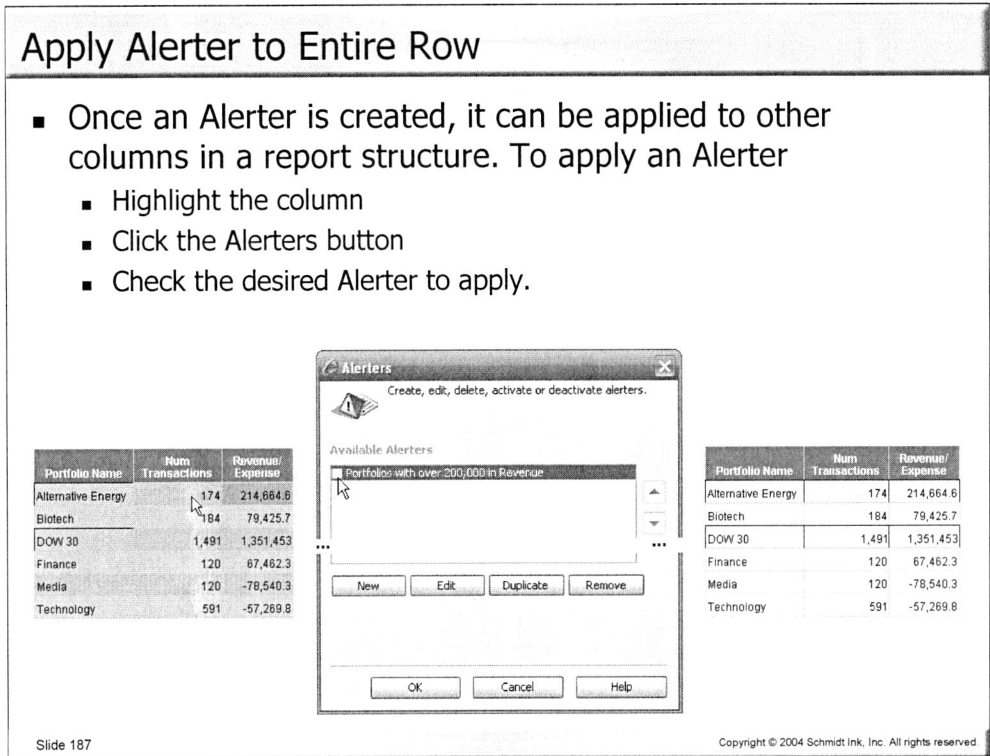

Apply Alerter to Entire Row
Once an Alerter is created, it can be applied to other columns in a report structure. To apply an Alerter
Highlight the column
Click the Alerters button
Check the desired Alerter to apply.
Alerters
Create, edit, delete, activate or deactivate alerters.
Available Alerters
Portfolios with over 200,000 in Revenue
New
Edit
Duplicate
Remove
OK
Cancel
Help
Portfolio Name
Num Transactions
Revenue/ Expense
Alternative Energy 174 214,664.6
Biotech 184 79,425.7
DOW 30 1,491 1,351,453
Finance 120 67,462.3
Media 120 -78,540.3
Technology 591 -57,269.8
Slide 187
Copyright © 2004 Schmidt Ink, Inc. All rights reserved.

We can apply an Alerter to a column after it has been created. We can also remove the effects of an Alerter by clearing the check in the option.

Apply an Alerter to the Entire Row

1. Click on any value in the Num Transactions column to select it.
2. Click the Alerters button in the Reporting toolbar.
3. Check the Alerter that we made in the previous exercise.
4. Click OK.
5. Repeat the above steps the Revenue/Expense column.

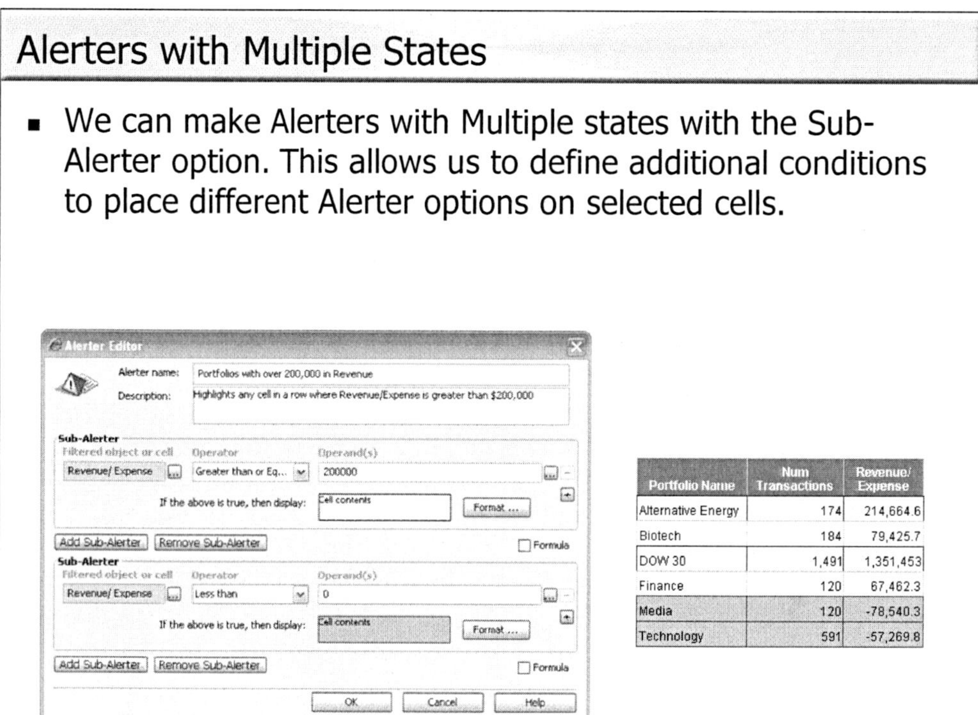

Portfolio Name	Num Transactions	Revenue/ Expense
Alternative Energy	174	214,664.6
Biotech	184	79,425.7
DOW 30	1,491	1,351,453
Finance	120	67,462.3
Media	120	-78,540.3
Technology	591	-57,269.8

Apply a Multiple State Alerter

1. Click on any value in the table from the previous exercise.
2. Click the Alerters toolbar button.
3. Click the Edit Button.
4. Click the Add Sub-Alerter button to add an additional Sub-Alerter section.
5. Enter the object for the Alerter comparison
    1. Click the small button the right of the Filtered Object or Cell field.
    2. Click on the Select Object or Variable entry on the pop-up menu.
    3. Select Revenue/ Expense from the Objects and Variables dialog.
    4. Click OK.
6. Select Less than or Equal to from the Operator drop-list.
7. Click in the Operand edit field and type 0.
8. Click the Format... button
    1. Set the font color to Default
    2. Set the Background color to light red
    3. Set the border color to red.
    4. Click OK
9. Click OK

Alerters Object Comparison

- In the previous two examples, we typed values into the Value 1 field. However, there are two other options
  - List of Values
    This option allows us to select from the values of the variable. This option works best with Dates and Text dimension values.
  - Variables...
    We can compare the variable to compare to other variables in the report.

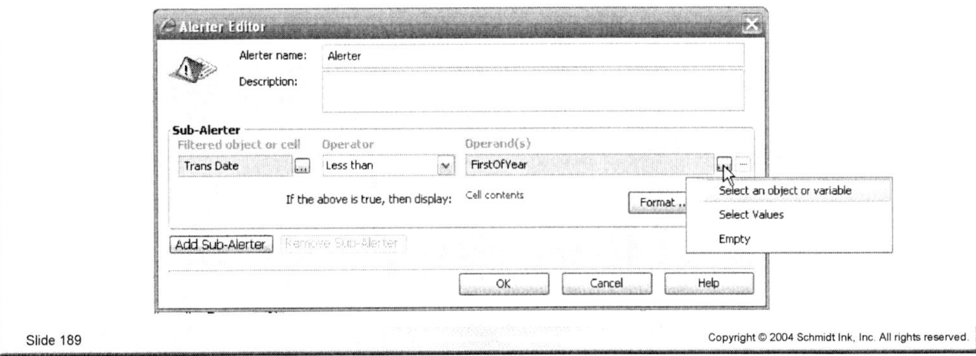

Slide 189

In this example, we create a variable to represent the first of the year. Then we use this variable in our Alerter comparison.

Use a Variable for Alerter Comparison

1) Create a report with Portfolio Name, Trans Date, Num Transactions, and Revenue/Expense.
2) Insert a column to the right of Revenue/Expense and click any cell in the column.
3) Click the Show/Hide Formula toolbar button on the Reporting toolbar.
4) Enter the following formula into the Formula bar
   = ToDate("1/1/2001"; "m/d/yyyy")
5) Click the Create Variable button
   1) Name the variable FirstOfYear
   2) Set the Qualification to Dimension
6) Remove the additional column.
7) Click on any value in the Trans Date column to select it.
8) Click the Alerters button.
9) Click the New button in the Alerters dialog.
10) Make sure that Trans Date is in the Filtered Object or Cell field.
11) Select Less than in the Operator drop-list.
12) Click the small button to the right of the Operand(s) field.
13) Click on Select an Object or Variable, select FirstOfYear from the object list.
14) Click the *Format...* button and set the desired format.
15) Click *OK*.

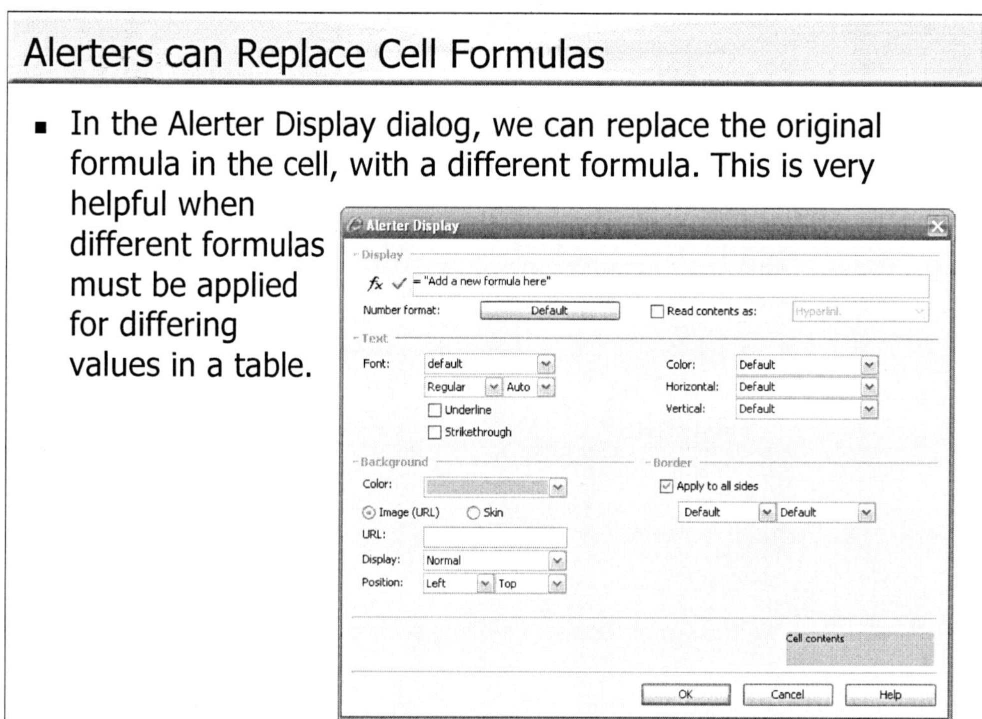

Earlier, we used the If function to alter formulas in a cell. The Alerter can do the same thing, but through the graphical interface of the Alerters dialog. It may be easier for some to use this interface. In addition, the new formulas can also be formatted differently, to help them stand out more.

Sorts, Filters, Ranks, and Alerters Summary

- This was an important chapter, because we learned how we can present the most important information in a report. We did this by
  - Sorting the information so that it is easy to find and maybe the most significant is on the top of the sort.
  - Placing filters that hide information that is not important.
  - Ranking the information and only showing the top or bottom of the rank.
  - Placing alerters that highlight information in a report.

Slide 191

We spend a lot of effort creating documents that give needed information to our report readers. This chapter showed us how to make sure the most important of this information is quickly discerned by these viewers of our reports.

Slide 192

Creating Documents with Web Intelligence XI

Query Techniques
(Java Report Panel)

Introduction

- Many reports require that we use data that is less than cooperative. Sometimes we are able to solve data problems in the report and sometimes we solve them in the Query Panel.
  - Earlier we used Dimension merging to link two queries in a report. This technique allows us to concatenate rows from the two providers.
  - Combination queries allow us to work vertically with data sets provided by two queries. They allow us to logically combine the rows into one data set.
    - We can Union, Intersect, and Minus the two queries
  - Many times we cannot isolate the data that is necessary in a report with conventional query filters. Sometimes, we will need to use a subquery to isolate the correct set for our reports.

Slide 194

Combination Queries

- BusinessObjects allows us to create three different types of combination queries. Each of these can only combine data from a single data provider. In the Venn diagrams below, the shaded color represents the data that is returned
  - Union Queries
    Use to combine similar sets of data.
  - Intersect Queries
    Use to intersect similar sets of data.
  - Minus Queries
    Use to subtract a similar set of data from another.

Slide 195

Combination queries are used to logically relate two sets of data. We can union, intersect, or minus the sets. BusinessObjects allows us to perform combination queries on sets of data created in a single data provider.

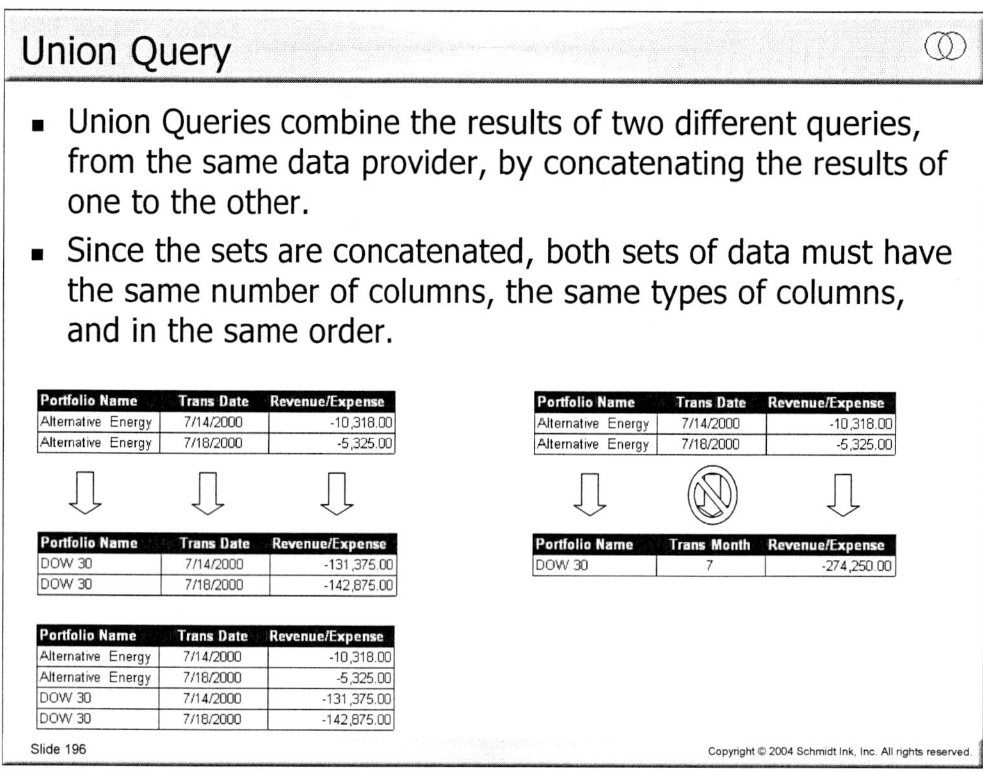
Union Query
Union Queries combine the results of two different queries, from the same data provider, by concatenating the results of one to the other.
Since the sets are concatenated, both sets of data must have the same number of columns, the same types of columns, and in the same order.
Portfolio Name
Trans Date
Revenue/Expense
Alternative Energy
7/14/2000
-10,318.00
Alternative Energy
7/18/2000
-5,325.00
DOW 30
7/14/2000
-131,375.00
DOW 30
7/18/2000
-142,875.00
Trans Month
7
-274,250.00
Slide 196
Copyright © 2004 Schmidt Ink, Inc. All rights reserved.

Union queries concatenate two sets of data from the same data provider. Union queries are often used to concatenate queries from two different contexts. They are also often used to concatenate data from the same context, but with different logical conditions.

Union queries are fast and efficient. Most databases perform union queries locally, thus eliminating the need for BusinessObjects to manage the data. This means that BusinessObjects sends a union query to the database and the database returns one unified set of data.

Therefore, once the data is returned to the document, it may be impossible to determine which query in the union returned a specific row. If this identification is needed, the report developer will usually place an object in each query that will uniquely identify the query from where the row came.

Since, the number, type, and order of columns must match in the queries, it may be necessary to include a NULL space holder for a column in one of the queries. Not all universes have these NULL objects, but it is highly recommended that they do. The SI EQUITY universe has these NULL objects.

Union Query Example

- Suppose that we had a manager switch portfolios after the first year. However, we need to get his/her two year total. We can use a union query to combine the first year results to the second year results.
- In this example, our manager worked in Biotech in 2000 and in Technology in 2001.

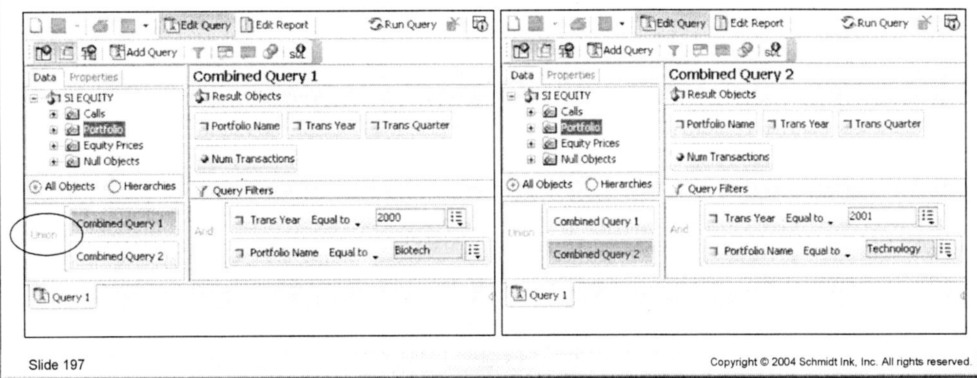

Slide 197

Create an Union Query

1. Create a query with *Portfolio Name*, *Trans Year*, *Trans Quarter*, and *Num Transactions*. Place the following conditions

   Trans Year = 2000

   Portfolio Name = Biotech

2. Click the *Combine Queries* button in the Query Panel

   Notice that the same objects occupy the Result Objects window, but the conditions are not duplicated. The redundant Result Objects are to remind you that the objects must match the first query's objects in type, number, and order.

3. Place the following conditions

   Trans Year = 2001

   Portfolio Name = Technology

4. Click the Run Query button.

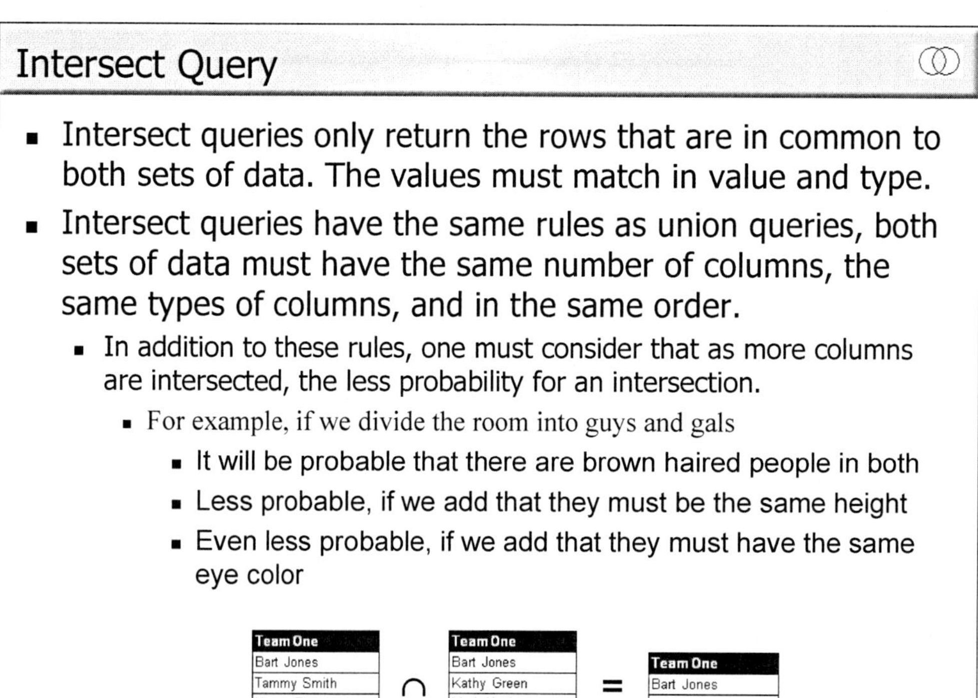

Intersect Query
Intersect queries only return the rows that are in common to both sets of data. The values must match in value and type.
Intersect queries have the same rules as union queries, both sets of data must have the same number of columns, the same types of columns, and in the same order.
In addition to these rules, one must consider that as more columns are intersected, the less probability for an intersection.
For example, if we divide the room into guys and gals
It will be probable that there are brown haired people in both
Less probable, if we add that they must be the same height
Even less probable, if we add that they must have the same eye color
Team One
Bart Jones
Tammy Smith
Henry Thomas
Wendy Willis
∩
Team One
Bart Jones
Kathy Green
Mark Hoyle
Wendy Willis
=
Team One
Bart Jones
Wendy Willis
Slide 198
Copyright © 2004 Schmidt Ink, Inc. All rights reserved.

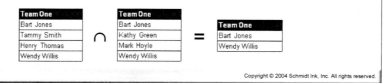

Intersect queries do allow us to intersect two queries. However, there use is limited, because the more columns we add - the less likely we are to get an intersection. Later, in this chapter, we will learn how to use a subquery to return the same information, but in more detail.

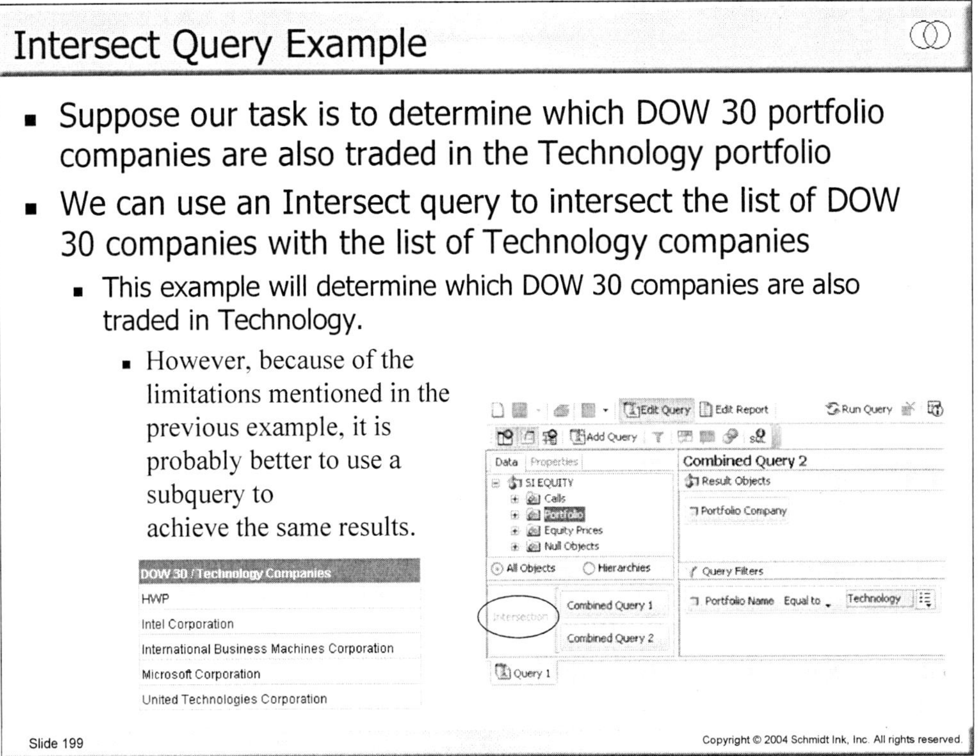
Intersect Query Example
Suppose our task is to determine which DOW 30 portfolio companies are also traded in the Technology portfolio
We can use an Intersect query to intersect the list of DOW 30 companies with the list of Technology companies
This example will determine which DOW 30 companies are also traded in Technology.
However, because of the limitations mentioned in the previous example, it is probably better to use a subquery to achieve the same results.
DOW 30 / Technology Companies
HWP
Intel Corporation
International Business Machines Corporation
Microsoft Corporation
United Technologies Corporation
Edit Query
Edit Report
Run Query
Add Query
Data
Properties
Combined Query 2
SI EQUITY
Calls
Portfolio
Equity Prices
Null Objects
Result Objects
Portfolio Company
All Objects
Hierarchies
Query Filters
Combined Query 1
Combined Query 2
Portfolio Name Equal to Technology
Query 1
Slide 199
Copyright © 2004 Schmidt Ink, Inc. All rights reserved.

Create and Intersection Query

1. Create a document with *Portfolio Company*. Place the following condition

   Portfolio Name = DOW 30

2. Click the *Combine Queries* button in the Query Panel

   To change the Union to an Intersection, double-click on the Union operator.

3. Place the following Query Filter

   Portfolio Name Equal to Technology

4. Click the Run Query button.

Minus Query

- Minus queries remove rows from a query, where they exactly match rows in a second query.
- Minus queries have the same rules as Intersect and Union queries. However, since the query subtracts the rows that are in common to the first query, order matters.
  - A U B = B U A, and A ∩ B = B ∩ A. However, A – B ≠ B - A

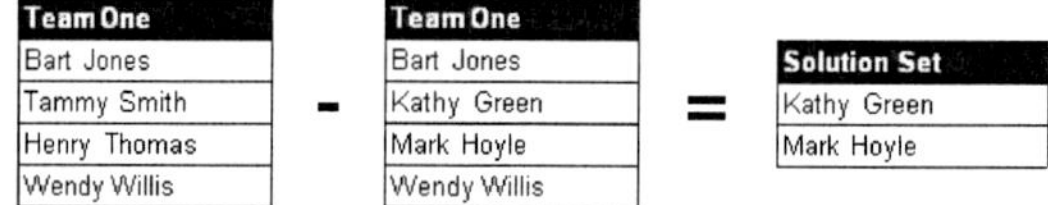

Slide 200

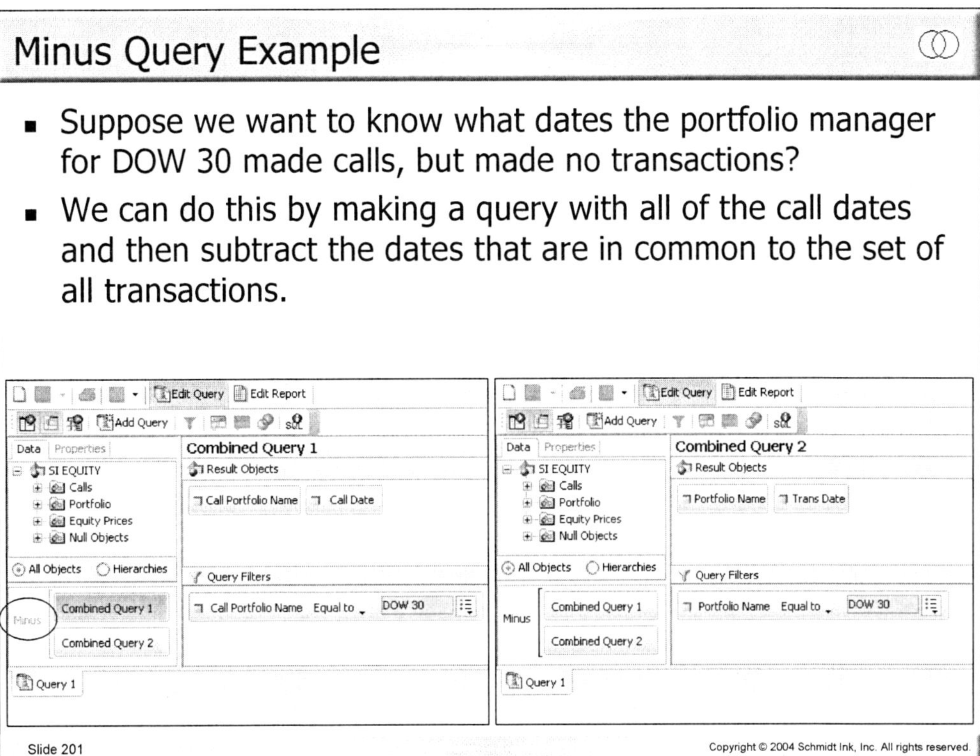

Create a Minus Query

1. Create a document with *Call Portfolio Name* and *Call Date.*
2. Place a Query Filter Call Portfolio Name Equal to DOW 30.
3. Click the *Combine Queries* button in the Query Panel

   To change the Union operator to a Minus operator, double-click it two times.
4. Remove both objects from the query and then place *Portfolio Name* and *Trans Date* in the *Result Objects* window
5. Place a Query Filter Portfolio Name Equal to DOW 30.
6. Click the Run Query button.

Combination Queries vs. Subqueries

- Of the three combination queries, two of them can usually be replaced with a subquery, which is more efficient and less restrictive.
  - These two combinations are Intersect and Minus. Subqueries allow us to intersect a single column, or minus rows based on a single column.
    - Since, we are using a single column for the comparison and not the entire row, we can add information that would not be allowed in the combination query.
- The Union query is the concatenation of two sets of data and therefore, cannot be replaced with a subquery that only constrains the amount of data returned to a document.

Slide 202

The In List Operator

- The condition below, will return rows to a query where Portfolio Ticker is AMD, AMAT or BOBJ.

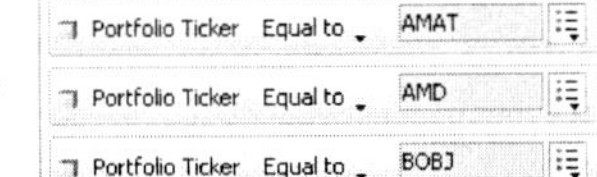

- The above condition will indeed return the desired results. However, since each condition is Or'ed together, it gets very awkward as more conditions are added. The condition below will return the same rows. However, it uses the InList operator and a list of values for an operand.
  - The list of values is easier to expand.
  - The InList operator makes the condition easier to work with, because the list is contained in a single condition statement

Slide 203

Both the Or and In List solutions in this example use a list. The first uses a list of query filters that are or'ed together. The second uses one query filter and a list of values. To expand the first solution, we have to add a new query filter and Or (add) it to the existing list. To expend the second, we simply add a value to the list of values in the existing query filter.

List of Values vs. Subqueries

- Many times a static list of values is what is needed for a condition.
  - Year InList 2000, 2001
  - Portfolio Name InList DOW 30, Alternative Energy, Media
- Subqueries create a dynamic list of values. In addition, this list has virtually no limitations on the number of entries.
- The query filter below, uses the In List operator. The operand list is a list of all Portfolio companies in the Technology portfolio.
  - This is much different from a static list of values, because it updates when new companies are added to the portfolio.

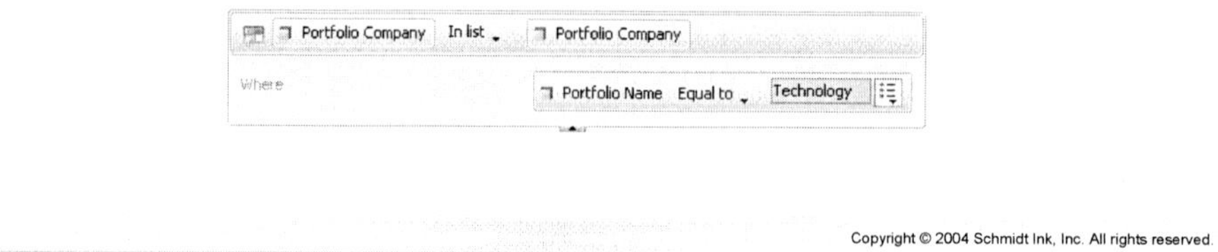

Slide 204

Using a list of values in the operand of a condition is great for hard-coded static values, such as years, states, or regions. However, if the list is dynamic and based on criteria within the database, it is very convenient to use a subquery to define the list.

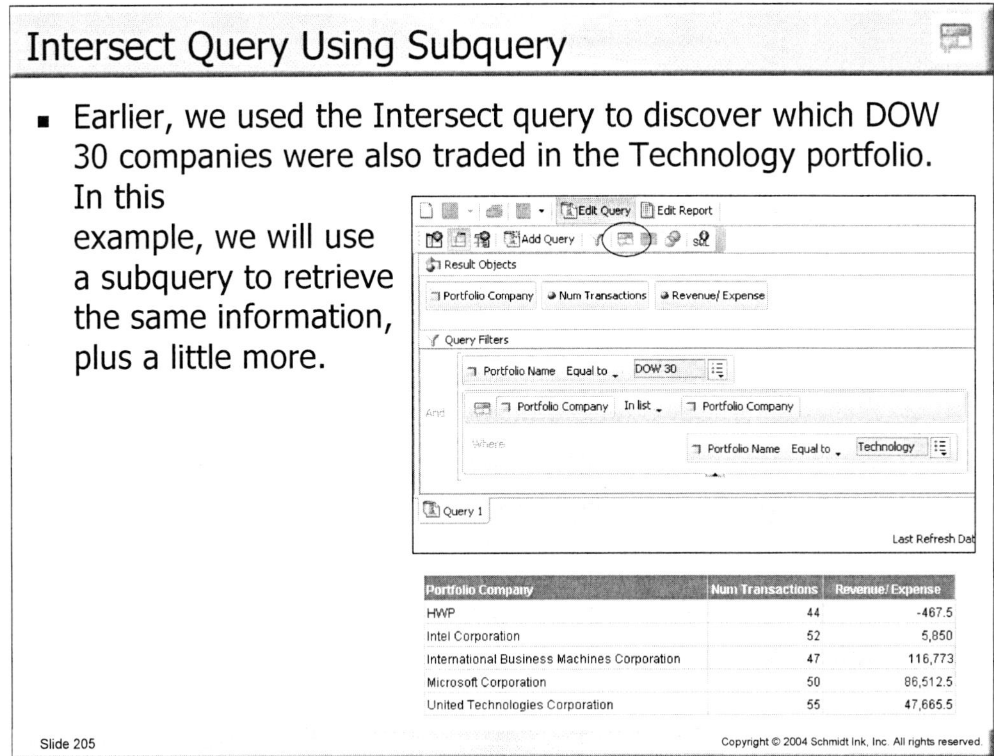
Intersect Query Using Subquery
Earlier, we used the Intersect query to discover which DOW 30 companies were also traded in the Technology portfolio. In this example, we will use a subquery to retrieve the same information, plus a little more.
Edit Query
Edit Report
Add Query
Result Objects
Portfolio Company
Num Transactions
Revenue/ Expense
Query Filters
Portfolio Name Equal to DOW 30
And
Portfolio Company In list Portfolio Company
Where
Portfolio Name Equal to Technology
Query 1
Slide 205
Copyright © 2004 Schmidt Ink, Inc. All rights reserved.

Portfolio Company	Num Transactions	Revenue/ Expense
HWP	44	-467.5
Intel Corporation	52	5,850
International Business Machines Corporation	47	116,773
Microsoft Corporation	50	86,512.5
United Technologies Corporation	55	47,665.5

Create an Intersect Query Using a Subquery

1. Create a query with *Portfolio Company*, *Num Transactions* and *Revenue/Expense*
2. Create a Query Filter *Portfolio Name* equal to DOW 30
3. Click the Add a subquery button.
4. If the two object spaces are blank, drop Portfolio Company into each of the spaces
   (Portfolio Company In List Portfolio Company)
5. Drag the Portfolio Name dimension to the filter area of the subquery
6. Select the Equal to operator
7. Select Technology from the list of values.
8. Click the Run Query button.

Since the outer query's columns don't have to match the subquery's columns, as in the Intersect query, we are able to retrieve much more information.

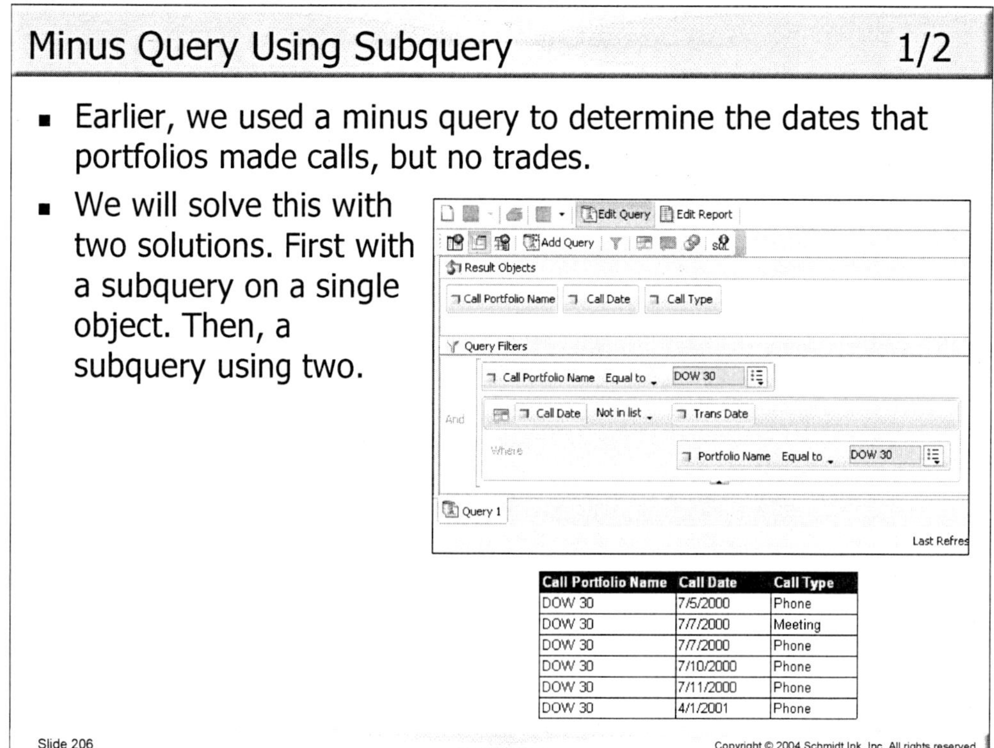

Call Portfolio Name	Call Date	Call Type
DOW 30	7/5/2000	Phone
DOW 30	7/7/2000	Meeting
DOW 30	7/7/2000	Phone
DOW 30	7/10/2000	Phone
DOW 30	7/11/2000	Phone
DOW 30	4/1/2001	Phone

Create a Minus Query Using a Subquery

1. Create a query with *Call Portfolio Name*, *Call Date* and *Call Type*
2. Create a condition *Call Portfolio Name* equal to DOW 30
3. Click the Add a subquery button.
4. Drop Call Date into the first object field of the subquery.
5. Drop Trans Date into the second field of the subquery.
6. Drag the Portfolio Name dimension to the filter area of the subquery
7. Select the Equal to operator
8. Select DOW 30 from the list of values.
9. Click the Run Query button.

Notice that both the subquery and the main query use the Query Filter Portfolio Name Equal to DOW 30. If we do not include it in the main query, then we will get all Portfolios that made calls and had no transactions on the dates that DOW 30 had transactions. Therefore, we need it in both the main query and the subquery.

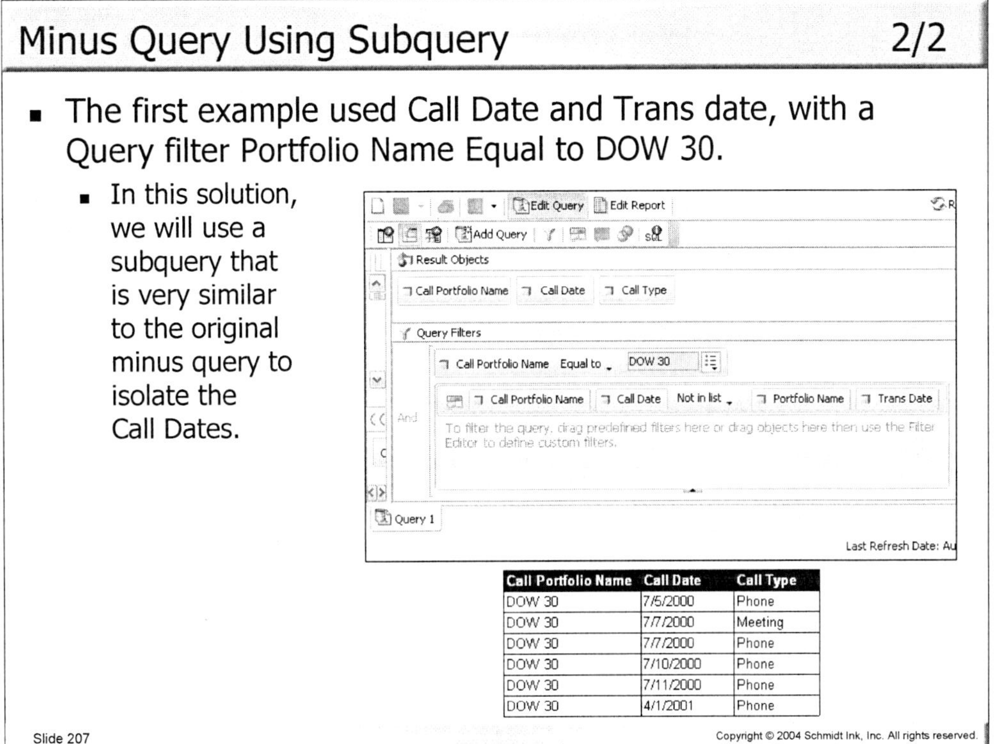
Minus Query Using Subquery
2/2
The first example used Call Date and Trans date, with a Query filter Portfolio Name Equal to DOW 30.
In this solution, we will use a subquery that is very similar to the original minus query to isolate the Call Dates.
Edit Query
Edit Report
Add Query
Result Objects
Call Portfolio Name
Call Date
Call Type
Query Filters
Call Portfolio Name Equal to DOW 30
Call Portfolio Name
Call Date
Not in list
Portfolio Name
Trans Date
And
To filter the query, drag predefined filters here or drag objects here then use the Filter Editor to define custom filters.
Query 1
Last Refresh Date:
Slide 207
Copyright © 2004 Schmidt Ink, Inc. All rights reserved.

Call Portfolio Name	Call Date	Call Type
DOW 30	7/5/2000	Phone
DOW 30	7/7/2000	Meeting
DOW 30	7/7/2000	Phone
DOW 30	7/10/2000	Phone
DOW 30	7/11/2000	Phone
DOW 30	4/1/2001	Phone

Multiple Object Subquery

1. Create a query with *Call Portfolio Name*, *Call Date* and *Call Type*
2. Create a condition *Call Portfolio Name* equal to DOW 30
3. Click the Add a subquery button.
4. Drop Call Portfolio Name and Call Date into the first object field of the subquery.
5. Drop Portfolio Name and Trans Date into the second field of the subquery.
6. Click the Run Query button.

Other Subquery Uses

- Subqueries can be used to return detail information using a summary condition
  - We can return the companies that generated more than $100,000. We can also return the trade dates, the portfolio names and, the number of transactions
- Subqueries also allow us to circumvent pre-established joins and contexts in a universe.
  - For example, suppose we wanted to know the closing prices and volumes for all of the equities in the Alternative Energy Portfolio. This query needs information from two different contexts in the SI Equity universe.

Slide 208

Details from Summary Condition 1/2

- In this example, we want to show the transaction details for the equities in our portfolios that generated more than $100,000 for the life of the fund.
- To do this, we will create a query with all of the details that we want to examine. This query will use a subquery to limit the equities to the companies that generated $100,000 or more.

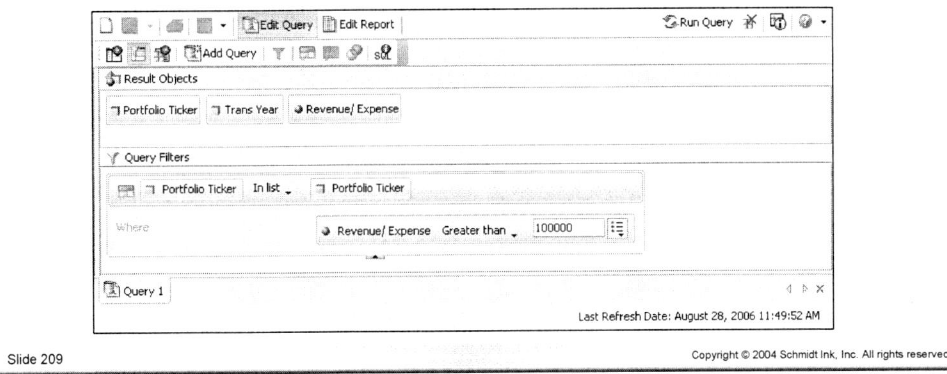

Slide 209

Creating a Summary Condition Using a Subquery

1. Create a query with *Portfolio Ticker*, *Trans Year*, and *Revenue/Expense*
2. Click the Add a subquery button.
3. Drop Portfolio Ticker into the first object field of the subquery.
4. Drop Portfolio Ticker into the second field of the subquery.
5. Drag the Revenue/ Expense measure to the filter area of the subquery
6. Select the Greater Than operator
7. Type 100000 into the operand edit field.
8. Click the Run Query button

Detail Info from Summary Condition 2/2

- Running this query will return detail information for the Portfolio Tickers in portfolios that generated more than $100,000.
- Other examples of detail information from a summary subquery include, but are definitely not limited to
  - Finding out who your top sales people are and who they sold products to.
  - Determining what instruments your top brokers are selling.
  - Returning the products that are sold in your most productive cities and a list of who is buying the products.

Portfolio Ticker	Trans Year	Revenue/Expense
AA	2000	-259,238
	2001	432,050
AA		**172,813**
IBM	2000	-567,825
	2001	698,848
IBM		**131,023**
MMM	2000	-380,363
	2001	577,350
MMM		**196,988**

Slide 210

Using a List of Values from a Different Context 1/2

- Suppose you want to know the closing prices for all of the stocks in a certain portfolio? This is difficult, because the closing prices are in a different context than the portfolio equities.
- The solution is to use a subquery to return a list of equities that are contained in a portfolio. A main query will use this list to limit the equities retuned by the main query.

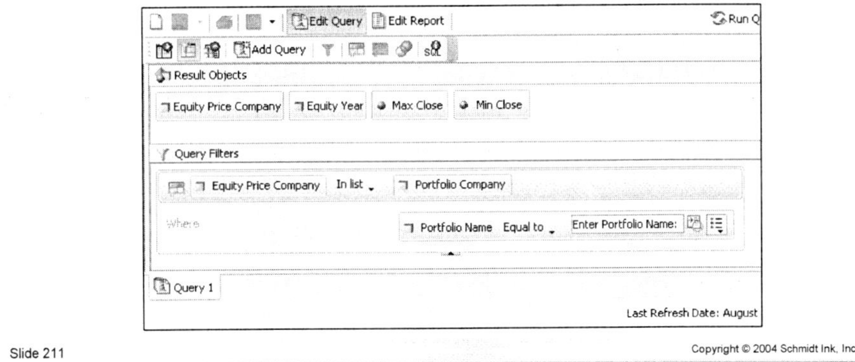

Slide 211

Using a List of Values From Another Context Using a Subquery

1. Create a query with *Equity Price Company*, *Equity Year*, *Max Close* and *Min Close*
2. Click the Add a subquery button.
3. Drop Equity Price Company into the first object field of the subquery.
4. Drop Portfolio Company into the second field of the subquery.
5. Drag the Portfolio Name dimension to the filter area of the subquery
6. Select the Equal to operator
7. Select the Prompt operand
8. Click the Run Query button
9. Select any Portfolio Name from the prompt dialog
10. Continue on next page…

Using a List of Values From a Different Context 2/2

- In this example we used a list of Portfolio Company Names from the Portfolio context to limit the companies in a query that returns the Max Close and Min Close for each company in the portfolio list.

Equity Price Company	Equity Year	Max Close	Min Close	Max - Min	% Move
AOL Time Warner Inc.	2000	63	34.8	28.2	81%
AOL Time Warner Inc.	2001	56.6	32.39	24.21	75%
Fox Entertainment Group, Inc.	2000	33	15.81	17.19	109%
Fox Entertainment Group, Inc.	2001	29.28	17.88	11.41	64%
Viacom Inc.	2000	74.63	44.94	29.69	66%
Viacom Inc.	2001	59.58	39.79	19.79	50%
Walt Disney Company	2000	42.5	26.44	16.06	61%
Walt Disney Company	2001	34.5	26.91	7.59	28%
Yahoo! Inc.	2000	139.81	25.63	114.19	446%
Yahoo! Inc.	2001	42.88	11.38	31.5	277%

Slide 212

1. Insert two columns to the right of *Min Close*
2. Insert the following formula into the first inserted column

   = [Max Close] – [Min Close]
3. Label the column

   Max - Min
4. Insert the following formula into the second inserted column

   = ([Max Close] - [Min Close]) / [Min Close]
5. Label the column

   % Move

Query Techniques Summary

- In this chapter
  - We learned how to retrieve only the data that is needed from a data provider. We learned how to isolate needed information using intersection and minus techniques. We also learned how to concatenate two queries using the Union query.
  - We also examined how to use subqueries to replace the Intersect and Minus combination queries. We found that the subquery allows us to return more details than the combination queries.

Slide 213

In most companies all of the data that we need for a document is usually available in the database or universe. The trick is getting it into our reports. This chapter introduced us to several query techniques that help us to isolate the data needed for our reports.

Slide 214
Copyright © 2004 Schmidt Ink, Inc. All rights reserved.

Schmidt Ink, Inc.

Creating Documents with Web Intelligence XI

Viewing Reports
(Interactive Viewing)

Introduction

- Now that we have used the Java Report Panel to create reports, we should look how others, and ourselves, may view the reports in the future.
- There are three options for viewing, although not all may be available on your installation or configuration.
- The available view formats are
  - HTML: This viewer has the fewest options, as the report is formatted in HTML. There is little interaction with the report.
  - Interactive: This format has the most options, as you can even alter the format of the report, add formulas, alter sorts, and so forth.
  - Portable Document Format: This format locks in the report so that viewers cannot alter it. People that are used to using Acrobat will appreciate this format, but they will not be able to alter the report.

Slide 216

Web Intelligence gives us three different viewing formats. Each one has it's own benefits. I have even worked in companies that view different report with different viewing formats. In this chapter, we will discuss the interactive viewer.

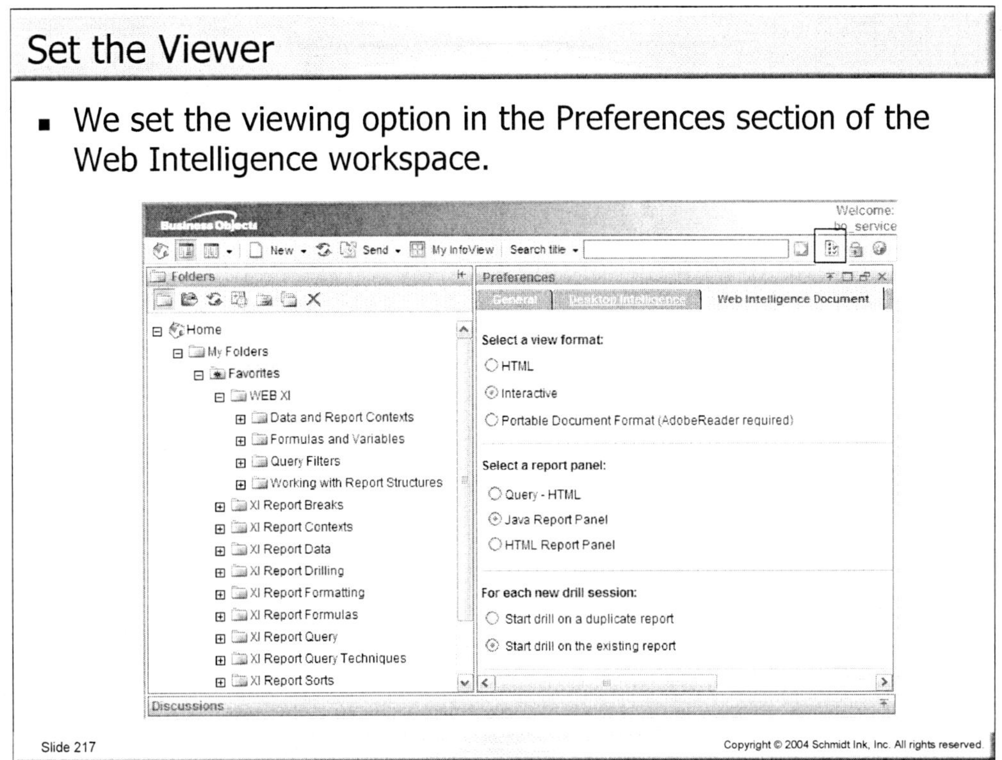

We set the viewing and editing platforms in the Preferences section of the Web Intelligence workspace. To get to the Preferences section, click on the Preferences button. Then, click the Web Intelligence tab to reveal the WEBI options.

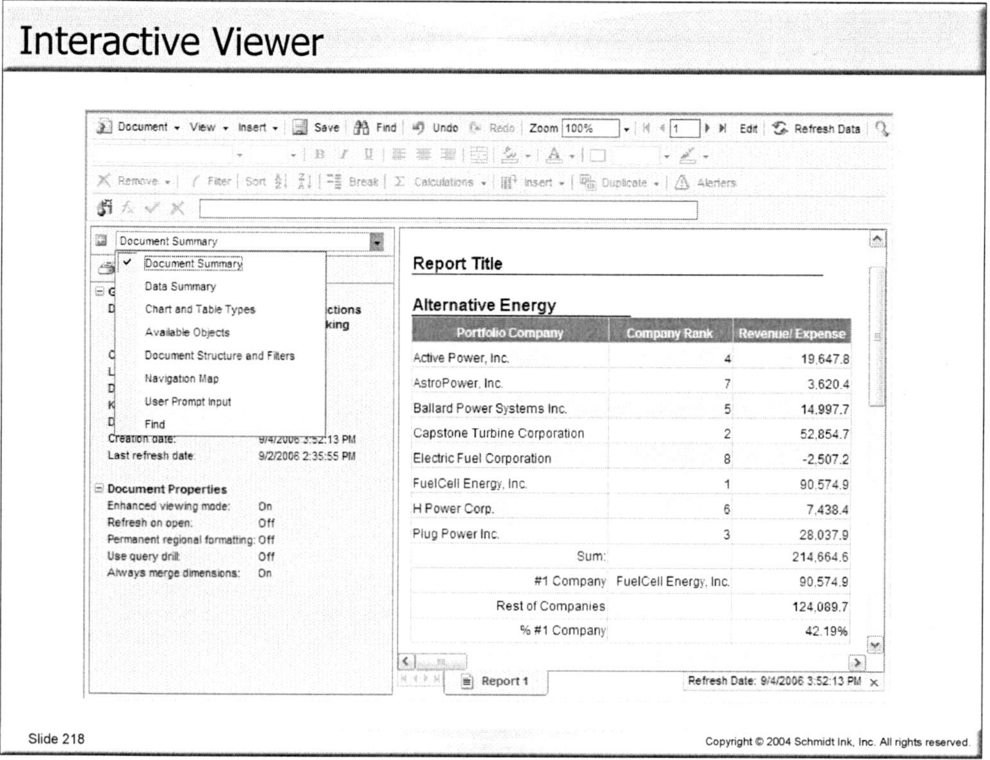

The Interactive viewer is very powerful and allows us to do many of the things that we could do in the Java Report Panel. It has several sections, which are:

- The Summary/Manager area along the left of the viewer. Here we can view the Document and Data Summary, use the Chart and Table templates, take advantage of the Available Objects, view all of the report structures in the document and place filters on any of them, use the navigation map, see the Prompt input, and find any text on the report.
- The Formatting, Report and Formula toolbars. These toolbars have much of the same functionality that we had in the Java Report Panel.
- The report viewing area. This is where we view the results of the report.

The Summary/Manager Area (Document Summary)

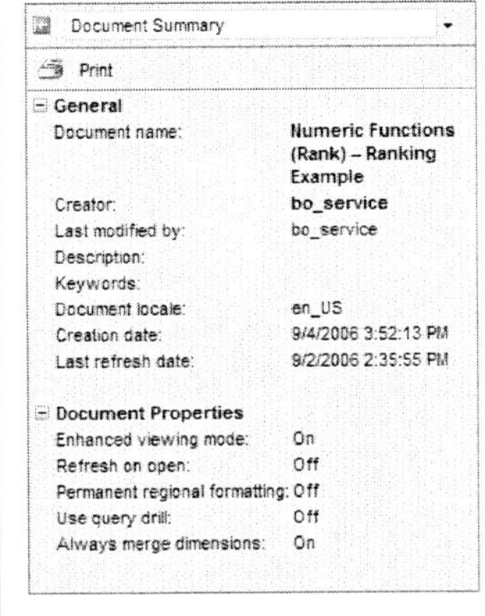

- The Document Summary displays information that helps us to identify a document. It also allows us to see the properties of the document.
- We can print the summary by clicking the Print button located near the top of the section.

Slide 219

The Document Summary displays information about the document.

The Summary/Manager Area (Data Summary)

- The Data Summary displays information about the data in a document. There are two sections: Data Source and Objects.
  - Data Source: This area displays the universe name, the execution time and the number of rows returned.
  - Objects: This area has three parts
    - Query Objects: These are objects that were selected in the query.
    - Document variables: Are variables that were created in the document.
    - Formulas: Are the formulas that are in the document.

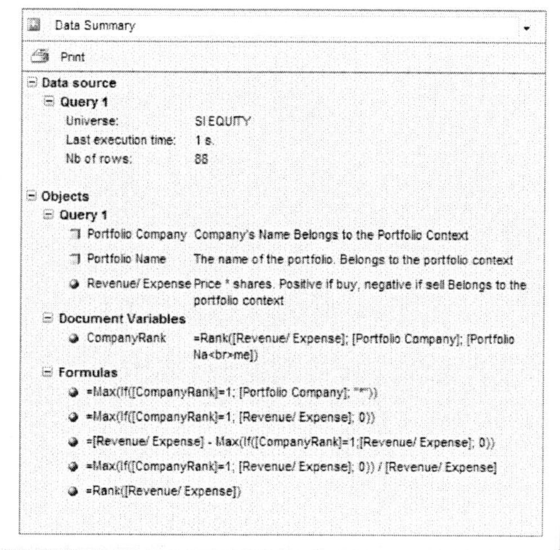

Slide 220

Data is very important to any document. The data determines much of a document's behavior and personality. Many companies have many universes, and some even have many universes with the same data. Sometimes, it is difficult to tell which universe has been used to create a document, which is why the universe is displayed in the Data Source section. Also, since Web Intelligence will roll-up or aggregate duplicate rows, it is possible for a document to display relatively few rows, while the query has returned many times more. Therefore, it is convenient that the number of rows is also displayed for us to view.

Reports don't always display all of the available objects in a document. In addition, many times object names are ambiguous and we are not sure what they represent. The Query section displays all of the objects that are available and the descriptions assigned to them by the Universe Designer.

Sometimes a variable does something, but we are not sure why. We may not trust the results or just want to see how it was defined. The Document Variables section list all of the document variables and the formulas used to create them.

Some reports have many formulas. Sometimes we modify formulas and then save the document. Sometimes, we need to replace measures with other measures. The older a document, chances are, the more times it has been modified. It is a good idea to browse through the formulas section to make sure that they are all still properly defined.

The Summary/Manager Area (Chart and Table Types)

- The Charts and Table Types section has all of the available templates for report structures. You can apply any of the templates by dragging a template from the Manager and dropping it on a report structure.

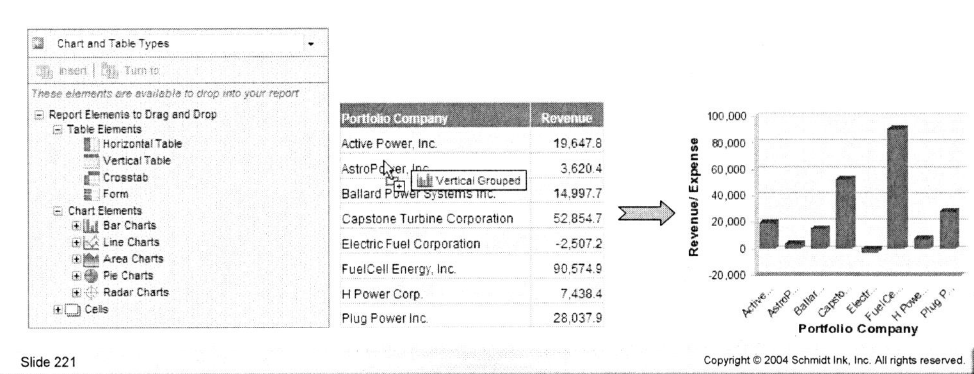

Portfolio Company	Revenue
Active Power, Inc.	19,647.8
AstroPower, Inc.	3,620.4
Ballard Power Systems Inc.	14,997.7
Capstone Turbine Corporation	52,854.7
Electric Fuel Corporation	-2,507.2
FuelCell Energy, Inc.	90,574.9
H Power Corp.	7,438.4
Plug Power Inc.	28,037.9

Slide 221

Sometimes it is just better to view a table as a chart or vice versa. The Chart and Table Types section allows us to apply templates to report structures. If you apply a template and you do not like the results, click the Undo button on the main toolbar.

The Summary/Manager Area (Available Objects)

- The Available Objects section is very similar to the Data Summary section. However, the objects in the Available Objects section can be dragged onto the report.
- A single dimension object can be dragged onto a report to create a section, a group of objects can be dragged onto the report to create a table, and a measure can be dragged into a table to create a new column.

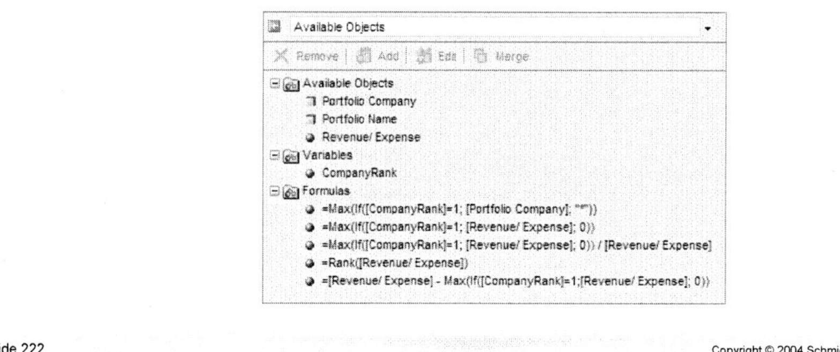

Slide 222

Create a Table with Available Objects

1. Open an existing Web Intelligence document.
2. Right-click on a report tab and select Insert to insert a blank document.
3. Select the Available Objects section from the drop-down in the Manager.
4. Select several objects in the section and drop them on to the blank report.
5. Select multiple objects

Document Structure and Filters

- The Document Structure and Filters section displays all of the structures in a document in a hierarchical fashion. It also displays any filters applied to the structure.
- Filters can be applied to report structures and report structures can be formatted.

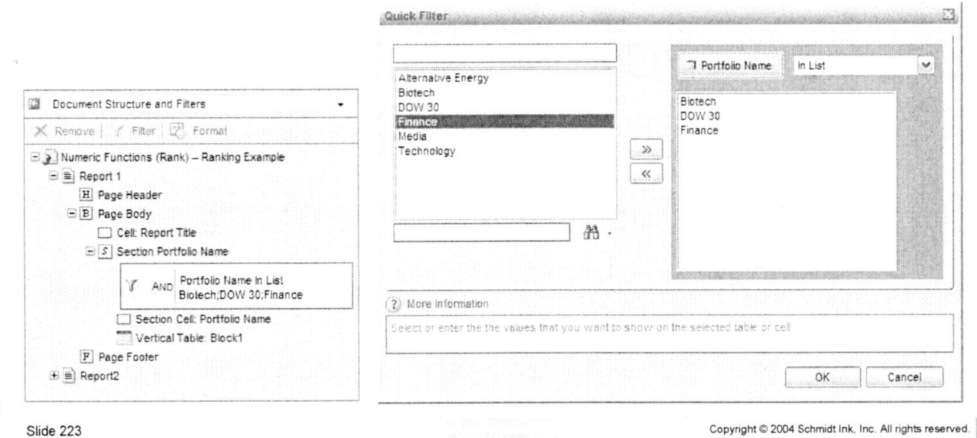

Most of the time, people are not worried about all of the structures in a report. They mostly care that all of the necessary information is visible and understandable. However, it is very important to see if any filters have been applied to report structures, because these filters effect the way a report behaves.

Apply a Quick Filter through the Document and Structures Section

1. Open an existing Web Intelligence document.
2. If there are no sections on the report, create one by dragging a dimension object from the Available Objects section.
3. Select the Document Structure and Filters section from the drop-down in the Manager.
4. Select the section structure and click the Filter button.
5. Select one or more of the available values.
6. Click the OK button.

Filters on Blocks or Reports

- Filters on Blocks and Reports are more complicated than filters placed on a single column or section cell, because there is usually more than one object.

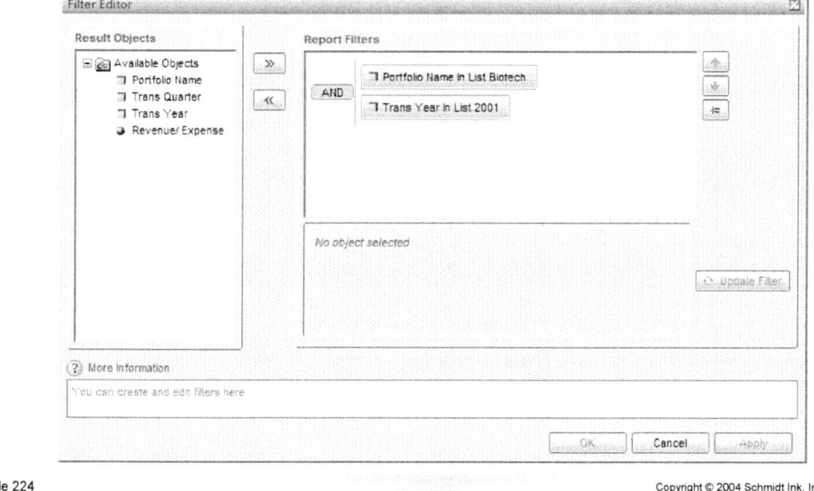

Slide 224

When a filter is placed on a report structure that has more than one object, the logic is more complicated. It is more complicated, because we must define how the logical statements are concatenated. In the example, the logic statements are concatenated with an And operator. Since we can build such complicated structures, we can create elegant filters that allow us to see just the data that we are interested in.

To change the And operator to an Or operator, just double-click on it. To nest or shift operators to the left, use the third button down, on the right, which is the Add Nested Filter button.

The Summary/Manager Area (Navigation Map)

- The Navigation Map works with sections in a report. The map allows you to click on a desired section in the map to navigate to that section on a report.
  - If the report that you are viewing does not have sections, then you can drag a dimension from the Available Objects and drop it on the report to create sections that are convenient to navigate.

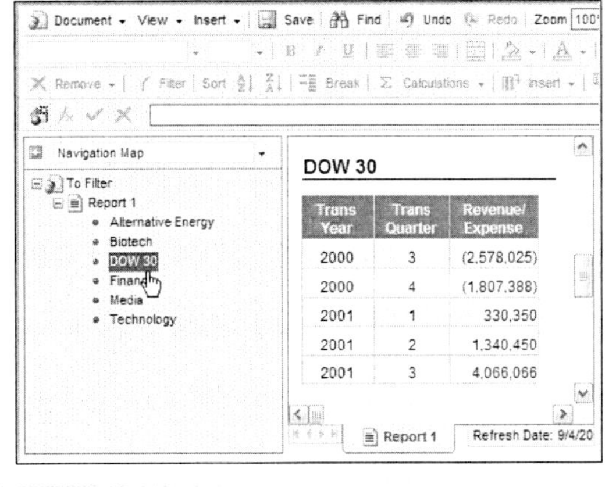

Slide 225

The Navigation Map makes it very convenient to find sections in a document.

Use the Navigation Map

1. Open an existing Web Intelligence document.
2. If there are no sections on the report, then drag a dimension from the Available Objects and drop it on the report to define sections.
3. Sect the Navigation Map section from the Manager drop list.
4. Click on the sections in the report to navigate to them.

The Summary/Manager Area (Find)

- The Find section allows us to locate text in a report. By default, you can enter any portion or case of the text. If you want to search for whole words, then select the option. There is also a Match Case option.

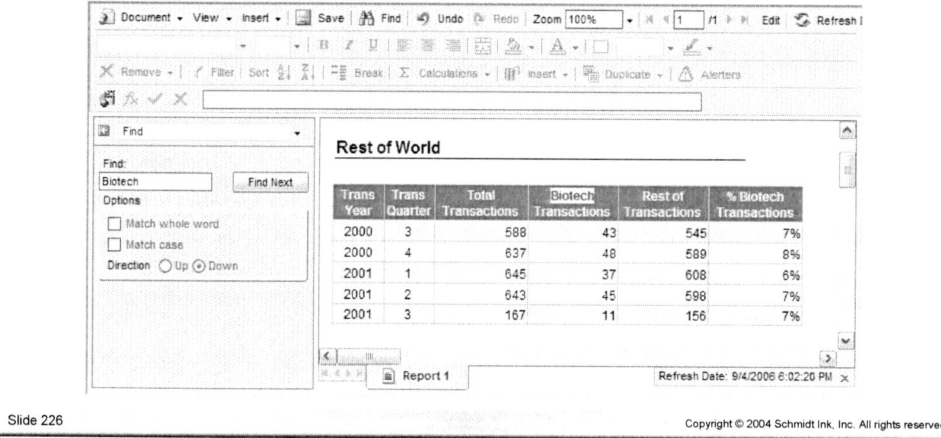

Trans Year	Trans Quarter	Total Transactions	Biotech Transactions	Rest of Transactions	% Biotech Transactions
2000	3	588	43	545	7%
2000	4	637	48	589	8%
2001	1	645	37	608	6%
2001	2	643	45	598	7%
2001	3	167	11	156	7%

Slide 226

To locate text in a document use the Find section of the Manager.

The Summary/Manager Area (User Prompt Input)

- Many documents prompt for input when they are refreshed. Most of the time the values entered in the prompt are displayed on the report. However, if they are not, then the values can be seen on the User Prompt Input section of the Manager.

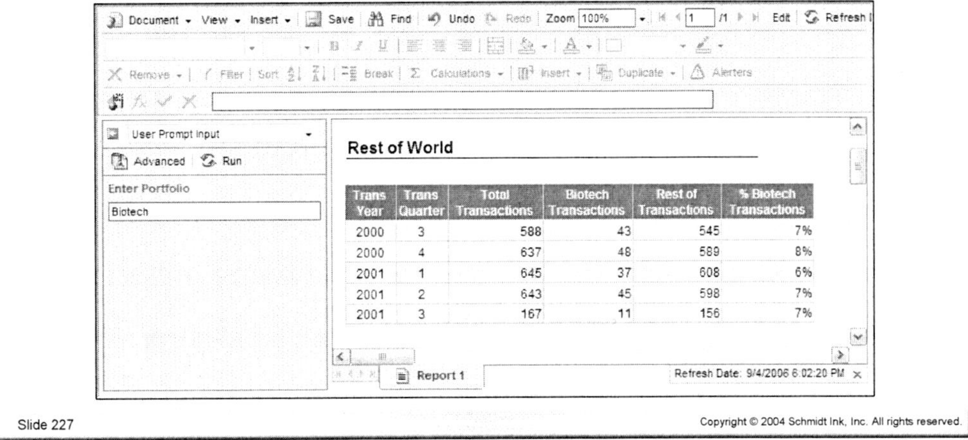

Trans Year	Trans Quarter	Total Transactions	Biotech Transactions	Rest of Transactions	% Biotech Transactions
2000	3	588	43	545	7%
2000	4	637	48	589	8%
2001	1	645	37	608	6%
2001	2	643	45	598	7%
2001	3	167	11	156	7%

Slide 227

Earlier, in this class, we created the Prompted Rest of World report. This report prompted for a portfolio name to compare with the rest of the portfolios. The report is displayed above and the prompted input is displayed in the User Prompt Input section.

The Formatting Toolbar

- The Formatting toolbar allows us to format the attributes of cells. We can...
  - Define the font and font size
  - Bold, italicize, and/or underline the contents of cells
  - Right, center, or left justify the contents of cells
  - Merge multiple cells into one larger cell
  - Fill a cell's background color
  - Assign a font color
  - Define borders, their width and color
- The formatting toolbar is very convenient, because it allows us to quickly format attributes in our reports.

Slide 228

The Formatting Toolbar allows us to quickly format attributes in a report.

The Report Toolbar

Remove | Filter | Sort | Break | Calculations | Insert | Duplicate | Alerters

- The Report toolbar allows us to modify reports. We can...
  - Remove a selected structure from a document.
  - Apply a filter to the report or selected structure.
  - Sort selected cells
  - Apply breaks to columns and/or Rows
  - Insert calculations
  - Insert Rows and/or columns
  - Duplicate report structures
  - Turn Alerters on and off
- The Report toolbar allows us to alter the way the information in a report is organized.

Slide 229

The Formula Toolbar

=[Portfolio Name]

- The Formula toolbar allows us to modify and add formulas to a report. We can...
  - Create and validate formulas.
  - Create document variables
- The Formula toolbar is very powerful, because it allows us to alter the information displayed in a report. It is the toolbar that can get you into the most trouble or get you the most praise.
- To use the Formula toolbar, you must understand formula syntax and know the available functions. These topics were covered in the Formulas, Variables And Various Functions chapter earlier in this manual.

Slide 230

The Formula toolbar is the only toolbar that can actually alter the information in a report. Therefore, it is very important that you understand formula syntax and the available functions, before you use it. To use it, you simply type a formula into the edit field and then press the [Enter] key.

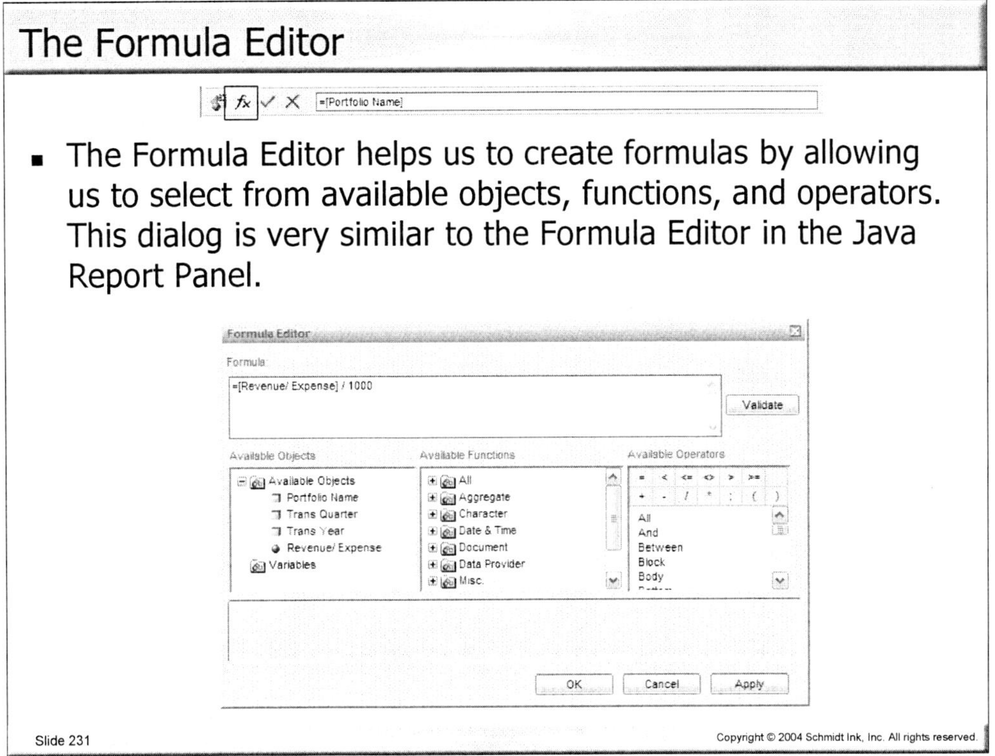

The Formula Editor helps us to create formulas with the proper syntax.

The Main Toolbar

- The Main toolbar allows us to...
  - Close, Edit, Save, and see a document's properties. We can save as WEBI, Excel, PDF, or CSV.
  - View different components of the environment, such as toolbars, the Manager, and the Status Bar. We can also view our documents in PDF mode, if we have the Adobe Reader Plug-in.
  - The Insert menu allows us to insert rows and columns, breaks, filters, and Calculations. It is redundant of the Report toolbar.
  - The Save button allows us to quickly save a document.
  - Find allows us to locate text in a report.
  - Undo and Redo for mistakes.
  - We can zoom in or out.
  - The page navigation allows us to navigate the pages in a report.
  - Edit launches the Java Report Panel, if this is our assigned editor.
  - Refresh Data refreshes the data in a document.

Slide 232

The main toolbar allows us to work with our document on the document level.

Viewing Reports Summary

- In this chapter
  - We found out that we can view our reports in an interactive environment. This environment is very user friendly and powerful.

Slide 233

End of Course

- If you have worked through all of the examples in this course and listened to all that the instructor had to say, then you must be very fatigued. However, you also must feel very accomplished, for you have done what most other report developers have not. You have explored most of the advance capabilities in BusinessObjects and your level of understanding has greatly increased.

- Thank you very much for taking this course and I hope that it will give you great confidence in BusinessObjects and other applications that may use similar logic and report structures.

Sincerely,
Robert D. Schmidt
RSchmidt@SchmidtInk.com

Slide 234